Metanarrative and the Environment

To meet the challenge of global environmental degradation activists have tackled clear and concrete problems such as carbon emissions and climate change, the ruination of ecosystems and habitat, the precipitous loss of biodiversity, and many other unhappy consequences of irresponsible human behaviour. However, all such efforts to manually correct the course of history have been dwarfed by the magnitude and heavy forward momentum of modern industrial society. In *Metanarrative and the Environment*, Stephen James Purdey argues that material approaches to the environmental crisis cannot succeed without the power of a legitimating discourse – a new metanarrative – which fundamentally changes the ideational landscape of human development. Dr. Purdey begins in Part I by establishing the pragmatics of our environmental predicament – its roots and responses to it. He focuses on the concept, definition, and key features of metanarrative, introducing the hegemonic story that now rules the contemporary global mindscape. Part II takes on the moral problematic more directly, encouraging the evolution of a new metanarrative by bringing our potential for agency in the face of danger into sharper relief. *Metanarrative and the Environment* is multidisciplinary, with a particular emphasis on the creative humanities. It will be of interest to undergraduate and graduate students alike, as well as environmental activists and academics looking for a new way forward.

Stephen James Purdey is an International Relations specialist (PhD, University of Toronto). His academic research focuses on the theoretical and normative attributes of systems of global governance, and on the practical evolution of new forms of global governance to meet current socio-ecological challenges. Now retired, Dr. Purdey worked for several years in the private sector, in Canadian federal politics, and with non-governmental organizations such as the United Nations Association, the World Federalists, and the Earth Council.

Routledge Research in Environmental Policy and Politics

Metanarrative and the Environment

A Story of Morality, Agency, and Governance

Stephen James Purdey

R Routledge
Taylor & Francis Group

NEW YORK AND LONDON

First published 2024
by Routledge
605 Third Avenue, New York, NY 10158

and by Routledge
4 Park Square, Milton Park, Abingdon, Oxon, OX14 4RN

Routledge is an imprint of the Taylor & Francis Group, an informa business

© 2024 Stephen James Purdey

ISBN: 978-1-032-64704-3 (hbk)
ISBN: 978-1-032-64705-0 (pbk)
ISBN: 978-1-032-64706-7 (ebk)

DOI: 10.4324/9781032647067

Typeset in Times New Roman
by Newgen Publishing UK

Contents

Part I

The Scope and Impact of Metanarrative

1 Metanarrative

An Introduction

On a wall
in a cave
these words are written:
We were here.

No longer a matter of speculation, but now a certainty; no longer an avoidable fiction, but now tomorrow's truth, the human adventure on Earth is about to take a very hard turn; the cost of our good fortune has come due. This book is about what it means to win everything, to lose everything, to start over from new beginnings. It is about the parlous relationship between people and planet, about how we have disrupted that relationship so violently and, for all intents and purposes, so permanently, that our tenure here is in serious jeopardy.

The ensuing discussion will not repeat the familiar litany of dismal facts and figures which support the allegation levied above; instead, I begin from the premise that what many are now calling a 'global environmental emergency' is real, palpable, imminent, and, I daresay, obvious and quite probably ruinous. Further, I adopt the view that our predicament is not solvable by technological know-how or policy innovations, valuable as those assets might be. Instead, the challenge of achieving planetary sustainability, insofar as that objective is achievable at all, presents humanity with a unique moral dilemma. William Ophuls makes this point abundantly clear in *Plato's Revenge*: "[T]he destruction of nature," he says, "is the consequence not of policy errors that can be remedied by smarter management, better technology, and stricter regulation, but rather of a catastrophic moral failure that demands a

DOI: 10.4324/9781032647067-2

radical shift in consciousness" (Ophuls 2011, 20). The Thomas Homer-Dixon is more hopeful about pushing the limits of human technical ingenuity but nonetheless gives voice to a similar opinion in his book *Commanding Hope*, noting that wrecking Earth would be "an unparalleled moral calamity, an unmitigated evil." What we are doing is "just wrong, plain and simple" (Homer-Dixon 2020, 350–1). He concludes that there is a need "to articulate clear moral and existential principles that position us all in a larger narrative of social purpose while giving guidance for what's right and fair" (Ibid., 341). While the moral dimension of global environmental degradation has certainly not been ignored by other professional academics (Leslie 1998; Gardiner 2011; Torres 2016), by responsible political leaders, or by theologians,[1] humanity writ large has not yet come to terms with the existential implications of 'catastrophic moral failure' nor has radical change been provoked by engaging the public consciousness in a transformative way. I intend to redress these deficiencies by focusing on two key problematics central to the challenge at hand, namely, morality and agency. To do so, this book will emulate the Aristotelian process of *phronesis*, that is, the pursuit of practical-moral knowledge instead of scientific or technical data.

Three themes emerge from these brief introductory comments which provide the foundations for all that follows. The first theme builds on Homer-Dixon's enigmatic reference to a 'larger narrative of social purpose.' As one might imagine, many parochial narratives that grant meaning to human affairs in various communities and cultures around the world already co-exist now. These local interpretive overlays take diverse forms such as origin stories or cultural histories and mythologies that provide context and depth to the quotidian lives of individuals. Green stories are also common today, stories which emphasize new ways of living together by highlighting local environmental successes and expounding key concepts such as holism, interdependence, stewardship, equity, and so forth, thought to be necessary, when scaled up, for the fruition of sustainable development on a planetary scale. But alongside this diversity of expression 'larger narratives' exist too, which envelope and, in a sense, override local stories; in fact, I will make the argument in what follows that one dominant story, one *metanarrative*, rules above all others, a hegemonic story so pervasive, potent, and persuasive that it literally defines the character and aspirations of the whole of human society and, significantly, determines which path we follow as we rush into the future. If this putative metanarrative is real, it must be deeply implicated in the causes of the currently dysfunctional relationship between people and planet, and in the global environmental emergency now upon us. And, if this is so, then presumably Homer-Dixon's still nascent 'larger narrative of social purpose' must be intended to redirect our evolutionary trajectory and to absorb and correct the moral deficiencies now embedded in today's incumbent metanarrative. The overarching objective of this book is to advance the evolution of this new story.

The second foundational theme introduced here draws on the notion of semiotics – the 'making of meaning' – which is an inherent feature and essential function of metanarrative. This quest for meaning may be the most telling dimension of the new story we will compose for and about ourselves. Here one encounters the ultimate questions with which the global narrative must grapple: what are we doing here; what, if anything, are we trying to accomplish? And, vexingly, why have we allowed ourselves to take ourselves, and countless other species, to the very brink of planetary eco-catastrophe? Lacking coherent answers, it has become clear that we as a species have lost touch with what is important – important for the physical continuity of ourselves and civilizations through time, for the healthy vitality of our spiritual being, and for Earth itself. The new metanarrative we seek will decisively shift the analytical terrain from 'how' questions – how can we achieve sustainability – to 'why' questions which probe more deeply into the metaphysical dimensions of life on Earth from which meaning emerges. We will undoubtedly continue to work in the trenches for concrete change but, I will argue, the battle for planetary sustainability, and for the survival of our species, will be won or lost in the ideational domain.

And, finally, the third theme which will thread its way through this book builds on the Aristotelian concept of *phronesis* mentioned earlier, namely, the pursuit of prudent moral knowledge that is grounded in and relevant to particular circumstances. Notwithstanding the prior importance of metaphysical issues, to be effective moral action must be commensurate with, and embedded in, the material realities of the contemporary world. On this account *phronesis* thus comprises both insight and practicality.

I will refer to the underlying dichotomy at play here as 'dualism,' understood generally in terms of the contrast between material (empirical) and immaterial (ideational) realities. This theme will recur consistently throughout the book, manifesting itself in diverse forms and finding its full expression in Part II where I discuss it under the heading of 'relationalism.' Dualism is also reflected in the structure of this book which I have divided into two parts. Part I will deal with the pragmatics of our environmental predicament – its roots and responses to it. Part II will take on the moral problematic more directly.

Continuing with the first of these three foundational themes, the remainder of Chapter 1 in Part I will focus on the concept, definition, and key features of metanarrative, introducing the hegemonic story that now rules the contemporary global mindscape. Chapter 2 covers a number of related issues. It opens with a short discussion about the importance of the creative humanities as a motive force for change, followed by a review of the basic elements of qualitative inquiry. I will then address in some detail the tension between one metanarrative and multiple local narratives, indicating that a single dominant story is not necessarily a benign construct. I will close this chapter (and foreshadow Part II) by introducing the 'moral turn' which this book recommends. Chapter 3 offers a discussion of the relationship between metanarrative and global governance, and examines the status of narration in contemporary postmodern culture. Chapter 4 presents a summary of the provenance of today's dominant metanarrative and, lastly, Chapter 5 will conclude Part I with a critique of that story and will address the practical requirements for sustainability which must be expressed in the new story. The specific contents of Part II, which takes up the 'moral turn' in earnest, will be listed in the opening commentary for that unit.

The story of us, envisioned as a metaphorical portrait of people on Earth, is still being written and this book will not finish it. My objective here is to encourage and to catalyse the conversation which must necessarily precede and inform our new trajectory to the future with the aim not only of urging a change of direction but, equally importantly, of arming ourselves with whatever courage it might take to survive the coming storm without turning violently on ourselves. Not every story, it must be said, has a happy ending. Some do not end well. When the promise of human excellence is exceeded by recklessness and self-regard, we have the unfortunate habit of capitulating to our worst nature, surrendering to hate and making enemies of each other. That fate would be deeply ignominious if all too familiar. It should be opposed in the material world with strength and modesty, and in the ideational world with whatever grace we can muster.

The Twin Pillars

There is no question that we humans are an extraordinary species. Unbothered by competition or predation, unconstrained by geography, endlessly adaptable to climate, seasons, and available foods, it is not surprising that, in our growing numbers, we have come to physically dominate the whole Earth. Blessed with a distinctive morphology, our rapid rise to supremacy was accelerated by another unique genetic advantage – an elite cognitive endowment that elevated us to unprecedented levels of perception, reason, and foresight with which we easily outranked and outperformed all other species. From these exceptional characteristics, both physical and intellectual, springs our ability, and perceived right, to do anything, to be anything, and to freely imagine a plentiful future in which all needs will be met, all wants satisfied.

And yet in spite of – or perhaps, as we shall see, because of – our exceptional capabilities we now find ourselves on the cusp of monumental changes which may dash those headstrong aspirations. Our global reach and brash enthusiasm have imposed, inadvertently or otherwise, major perturbations on land, air, and water setting in motion pervasive, long-lasting changes to how Earth works. So massive have these cumulative disturbances to the planet been that we are now entering a new geological interval, marking an end to the familiar environmental stability of the preceding 12,000 years. The aptly named Anthropocene epoch,[2] displacing the earlier Holocene during which humanity rose to pre-eminence among all species, acknowledges (albeit as a self-portrait) the dominion of human life on Earth, and it underscores the fact that we have wilfully taken ownership of our own destiny. It promises, however, no assurance whatever that we have the insight, the competence, or the courage to manage that destiny, or to guarantee a future which supports even the barest of necessities. We have arrived, in other words, at a crucial historical juncture which counterposes spectacular success with the real possibility of equally spectacular failure – a juncture which, without due care, will see conceit turn quickly to tragedy.

The truth of the matter is that our material success as a species has not been matched by an equally robust evolution of empathic care for our natural surroundings, by a reflexive appreciation of our own power and limitations, or by an emergent sense of adult responsibility. This imbalance obscures the fact that we are both Earthbound and transcendent beings, alive to the mysteries of the universe yet grounded in a material physicality. The latter provides a stream of tangible benefits necessary for life and prosperity; the former adds the possibility of evaluation, betterment, and purpose to the human experience. These two features of our existence should be harmonious, evolving together, forming the twin pillars of a resilient human community on Earth – but our extravagant success has made us arrogant, imperiously dismissing any notion of accountability to each other, to the planet, or to a loftier reality. Corrective action is now necessary to restore a balanced reciprocity between these two pillars, between the material and non-material worlds inhabited by individual people, and by human society writ large.

The process of reconciliation begins necessarily by acknowledging the obvious, that is, by facing directly the physical damage we have done to Earth's biosphere, the extremity of which now threatens our lives. We have altered the composition of the atmosphere and ocean with callow disregard for and little comprehension of the consequences. Ice fields, fresh water systems, soils, forest cover, species habitat, nutrient cycles, and much more have all proven vulnerable to clumsy human interventions, interventions that have now provoked a precipitous degradation of the planetary biosphere and a rising danger to our long-term well-being. This frank assessment foregrounds the material aspect of that threat – but it simultaneously alerts us to the ideational dimension of our relationship with the planet, now conspicuous as an uneasy apprehension rippling around the world that something is amiss, that danger is nearby, that our families and futures are not safe. Impelled by persistent media coverage of environmental damage, and often by the personal experience of it, this apprehension is expressed in a plethora of popular and professional publications, in governmental and private sector initiatives, and in widespread public demonstrations, all intended to emphasize and advocate solutions for what is generally (but by no means exclusively) perceived to be a global environmental emergency.

Solutions rising from public nervousness circle predominantly around core issues of mitigation and adaptation; typically, they address the practical need to lessen our environmental impact and adapt to changes already upon us. These are important but a growing subset of activists and analysts are now beginning to focus less on the nuts and bolts of remediation and more on the ideational underpinnings of our current environmental predicament. Here we find, for example, efforts to uncover hegemonic worldviews which inform and legitimize humanity's disruptive relationship with Earth,[3] and to expose the cultural narratives which convey those views through time and space. The objective of this effort is to better understand the provenance and social impact of these inappropriate beliefs, values, and stories, and to supplant them with a public mindset more conducive to planetary sustainability.

Worldviews, narratives, and the like are referred to as forms of 'discourse' in the academic literature. Important as these are, however, they are often regarded as too ephemeral to grapple with, too far removed from the actual labour of environmental protection. Practical people have chosen instead to tackle tangible problems, working to mitigate the worst effects of profligate energy consumption and climate change, of global industrialism and the ruination of ecosystems, of misbegotten wealth, widespread poverty, and many other unhappy consequences of irresponsible human behaviour. But this preference falls prey to the problem identified at the beginning, namely, that the balance between our transcendent and material inclinations has been lost, with too much emphasis on the latter. Without question, the pragmatic, hands-on approach to change is indispensable but the metrics are abundantly clear. To date, all such efforts to manually correct the course of history have been dwarfed by the magnitude and heavy forward momentum of modern industrial society. Business as usual proceeds as usual, and the costs continue to rise. The practical/material approach cannot succeed without the power of a new story which will fundamentally change the ideational landscape of human development.

This allusion to an 'ideational landscape of human development' refers to that stratum of thought inhabited by shared ideas, beliefs, and values. Clearly, this nebulous, multifarious landscape (sometimes called a noösphere after Pierre Teilhard de Chardin[4]) presents a formidable challenge across all three of the standard metrics of social science: what is it (ontology), how can we learn more about it (methodology), and what knowledge claims might be revealed by it (epistemology). Answers to these questions will be proffered in the following but, for the moment, I simply confirm that this ideational stratum must inevitably be a complex phenomenon. It is diverse in terms of range and types of concepts included, multi-layered, and, above all, dynamic, constantly changing in form, composition, and substance, all of which render it a profoundly important but poorly understood feature of the modern human experience. Helpfully, however, there appear to be major threads and themes running through it that bring a degree of order to the cacophony of competing ideas, images, and intentions which enliven the transcendent dimension of our shared experiences. I have already suggested that one such theme overshadows all others, a dominant theme which not only informs but concretely determines the developmental trajectory – that is, the direction chosen and the objectives sought – of the human population on Earth at large. I referred to this formative entity as a metanarrative. Because this conjecture bears significantly on the central problematic of this book, namely, how better to understand and, with good fortune, to resolve the global environmental peril now at our doorstep, several key features of it might be posited:

- There is now in place a dominant metanarrative (or, in colloquial terms, a 'story') called Progress and Prosperity which defines and determines the ideational contours of contemporary human life on Earth.
- That story is dysfunctional with respect to global ecological stability and to the long-run viability of the relationship between people and planet.
- Granting, however, that neither progress nor prosperity is inherently malign, but instead is disruptive only when coupled with a naïve exuberance and a licentious notion of freedom, those same objectives should be subsumed within a more mature story which properly engages an appropriate set of contextual parameters such as temperance, sufficiency, and a new appreciation of nature's realities.

- In keeping with my objective to restore a balanced reciprocity between the material and non-material worlds we inhabit – between the twin pillars of a resilient human community on Earth – the new parameters just recommended will comprise both phenomenal (material/experiential) and noumenal (ideational/normative) considerations.

- And finally, it will be a key and central assertion of this book that a reconsideration of the latter (noumenal) domain is of prior concern, vital to the resolution of the environmental challenge now confronting our species; specifically, that this challenge is at bottom a moral problem access to which will require a unique set of tools and a qualitative methodology capable of probing – and, indeed, changing – those ideational elements on the felicity of which we understand ourselves, and direct ourselves forward.

Metanarrative: Definition and Discussion

Known formally as a discourse, metanarrative is a story that has permeated the public domain and become an integral part of the social fabric. It can be defined as "a shared way of apprehending the world" resting on a set of assumptions, judgements, and contentions (Dryzec 1997, 8) or, put differently, "a matrix of social practices that gives meaning to the way that people understand themselves and their behaviour. A discourse ... generates the categories of meaning by which reality can be understood and explained" (George 1994, 29–30). Tangibly, it is the suite of beliefs, values, collective intentions, and ethical propensities which emerge from that dialogue, providing a pattern, direction, and an interpretation of events and practices that give meaning to the human experience.

A conversation among people about the performance of their local economy, for example, may be termed a 'capitalist discourse' insofar as it identifies roles (employers and employees), expectations (money for service), and norms (fair competition) associated with profit-oriented economic behaviour. A common understanding among participants in the conversation reinforces the normality, propriety, and desirability of its key tenets. This understanding facilitates communication which in turn increases the incentive for others to adopt it as well. In this way a discourse informs and is formative of social relations even as it offers a shared sense of meaning about the world around.

Deeply embedded and almost invisible, a metanarrative is about who we are and what we want and as such it influences profoundly what we believe to be true and what we understand to be right. And, it is dynamic; it evolves over time. The manner and direction of that evolution are critical to the quest to achieve planetary sustainability. Fortuitously, the non-material composition of ethics, values, expectations, and the like makes these social constructs susceptible to swift evolution unlike, for example, the mechanical challenge of re-tooling the world's energy infrastructure. This characteristic presents a unique strategic advantage in the fight for change.

A discourse may be comprehended under a variety of names – a story, an ideology, a worldview, a cosmology, a social episteme, a secular religion, or an ideational superstructure – and I will have occasion to refer to all of these from time to time. However, for the sake of simplicity and clarity, I will prefer the term 'metanarrative' in this book, by which I mean an overarching story reflective of and intrinsic to human society on Earth as a whole, and which defines the contours of the public mindscape. It is no exaggeration to assert at the outset that a metanarrative is, in effect, the command-and-control function for our collective behaviour. In addition to shaping social relations and expressing the aspirations we share, it informs the choices we make and, ultimately, determines how we manage the relationship between people and planet.

To be clear, however, metanarrative is not an immutable truth or a rulebook. It is not a bucolic fantasy, a cautionary tale, an epic myth, or a sacred text – although it may on occasion share features with all those things. At bottom, it is the conversation we have among ourselves about ourselves, and about our place in the cosmos. And, as with any conversation, its mood, content, and direction can change when new or contrary ideas are interjected and engaged. Given the pervasiveness and deep roots of the incumbent metanarrative, however, any such engagement will inevitably be disruptive and jarring. When ideas clash, when stories collide, the result is a new amalgam of beliefs and values, and a new direction forward. The shared process of creating a new story will be uniquely challenging – historically, intellectually, and morally – but, by equal measure, profoundly stimulating as it calls to question the deepest meaning of our species' tenure on Earth.

Discussion

Today, sociopolitical and economic macro-management of the development trajectory of human society on Earth is informed by the story of Progress and Prosperity, the provenance of which I take up in Chapter 4. The cultural critique on offer here is intended, essentially, to interrogate the merits of this story. It begins with a broad, top-down view of human affairs from which macro-trends of our collective behaviour become visible, especially the dysfunctional relationship between people and planet and the attendant global environmental crisis now upon us. Because progress and prosperity are the defining themes of the incumbent metanarrative, and because those themes both carry strong economic implications, the critique of this story falls properly under the heading of international political economy (IPE).

A caveat is required here, however, acknowledging that other (competing or complementary) perspectives are also available for scrutiny. Depending upon one's preferred analytical lens, other macro-trends come into view which exert their own brand of directive influence on our collective behaviour. A sociological/ Marxian approach, for example, shares some features with my own insofar as it recognizes that dominant ideas are structurally important to social relations. Instead of highlighting the source, content, and impact of those ideas, however, Marxism focuses instead on the seemingly natural tendency of human communities everywhere to organize themselves hierarchically into 'classes' – into leaders and led, rulers and ruled – a view which brings to focus elite power and privilege and the oppressive economic relations which those structures enable.[5] Ideas are relegated to a conditioning role even as Marxist analysis quite rightly exposes a transnational managerial class and the rapaciously consumptive power of capitalism which, in the modern context, is often fingered as the *bête noire* of environmental degradation, as the culprit in need of reform and sociopolitical oversight. I will have more to say about this in Chapter 4 where I subsume capitalist discourse into a broader package of cultural/ideational change.

A more visceral approach to understanding macro-human behaviour, the exercise of power, and, specifically, the destruction of nature, showcases our species' genetically programmed focus on the exigencies of the here and now, and on our atavistic insensitivity to anything not suited to survival. Our hard-wired psychology fosters short-term thinking, the immediate satisfaction of needs and impulses, and, importantly, a disdain for, indeed fear of, the 'other,' that is, any competitive humanoid group which may contest access to territory and/or vital resources. Notwithstanding the fact that some features of primordial human psychology such as altruism and self-sacrifice, foresight, care, and aesthetic expression are also thought to be inherent to the human psyche and available to be deployed by those so inclined, they are evidently easily overrun by the brute aggression and mindlessness which seem to be endemic features of the human animal, even today. Indeed, the ugliness of today's war-torn international landscape and the wanton destruction/consumption of massive swathes of Earth's natural capital speak plainly to this unseemly behaviour. This 'realist' point of view is very much in the mainstream of contemporary political science, providing academic cover for self-interest, power-seeking, and bellicose behaviour. It rests ultimately upon an unflattering portrayal of human nature, even as it offers a persuasive explanation for the causal forces that have led us so compulsively and rapidly to the very brink of global eco-catastrophe.

Finally, a radically different but structurally similar point of view derives from the feminist perspective which names patriarchy as the primordial source of hierarchical stratification of societies around the world. More like a social pathology, patriarchy features the systemic domination of male over female, a pattern which extends from the domestic hearth all the way to global politics and international economic relations. As a generic form of imposed power, patriarchy is a pervasive fact of life which colours and distorts virtually all aspects of economic, political, social, and cultural relations around the world, and of course its built-in hierarchical structure – especially its proclivity to suppress, dominate, and control – is deeply implicated in the maltreatment of nature (Zimmerman 1987). I leave this important line of interrogation to the competency of others.

This quick review of alternative modes of analysis is not intended to be comprehensive; rather, it serves to indicate that an IPE-based metanarrative is not the only way to grapple with formative trends which shape and, in some cases, determine humanity's macro-behaviour. Apart from the feminist perspective, I will have occasion in the following to draw on some of these alternative modes with regard to the problem of global environmental degradation as I develop my own analysis which, as mentioned, might best be called a cultural critique. I prefer the cultural/ metanarrative approach for four reasons. First, it focuses squarely on the ideational domain of our shared experience, the evolution of which has lagged far behind its materialist counterpart to the overall detriment of human society on Earth. Second, as indicated earlier, I intend to argue that the global environmental crisis is at bottom a moral dilemma, the resolution of which must necessarily begin from a critical scrutiny of the normative/noumenal – that is, the ideational – domain. Third, hegemonic metanarratives tend to be all-inclusive, a feature which makes them germane (whether perceived or not) to the lives and living conditions of people everywhere, regardless of gender, status, or station. And finally, unlike other approaches, metanarrative is closely tied to the actual structures and processes of global governance *per se* which provide (whether effectively or not) the sociopolitical oversight that determines the overall evolutionary trajectory of human society; in a word, metanarrative is policy-relevant, finding expression around the world today in the relentless pursuit of Progress and Prosperity.

Notes

1 Thomas Berry's prolific work is notable here. See, for example, Berry (2000).
2 Neither the International Commission on Stratigraphy nor the International Union of Geological Sciences has formally approved this term as a subdivision of geological time, though it is by now in common usage across several disciplines.
3 A common target for widespread opprobrium in this regard is capitalist ideology. See, for example, Naomi Klein's popular critique (2014).
4 For more on this, see www.organism.earth/library/document/format ion-of-the-noosphere.
5 These familiar themes of Marxist thought tend to overshadow his understanding of and contributions to the study of the metabolic relationship between society and nature, now an important part of the

environmental literature called Marxist Ecology. For a comprehensive summary of this field, see John Bellamy Foster (2015).

References

Berry, Thomas. 2000. *The Great Work: Our Way into the Future.* New York: Crown Publishing Group.

Dryzec, John S. 1997. *The Politics of the Earth: Environmental Discourses.* Oxford: Oxford University Press.

Foster, John Bellamy. 2015. "Marxism and Ecology: Common Fonts of a Great Transition." *The Monthly Review* 67, no. 7 (December). https://monthlyreview.org/2015/12/01/marxism-and-ecology/

Gardiner, Stephen M. 2011. *The Perfect Moral Storm: The Ethical Tragedy of Climate Change.* New York: Oxford University Press, Inc.

George, Jim. 1994. *Discourse of Global Politics: A Critical (Re) Introduction to International Relations.* Boulder, CO: Lynne Rienner Publishers.

Homer-Dixon, Thomas. 2020. *Commanding Hope: The Power We Have to Renew a World in Peril.* Toronto: Alfred A. Knopf Canada.

Klein, Naomi. 2014. *This Changes Everything: Capitalism vs. the Climate.* Toronto: Alfred A. Knopf Canada.

Leslie, John. 1998. *The End of the World: The Science and Ethics of Human Extinction.* New York: Routledge.

Ophuls, William. 2011. *Plato's Revenge: Politics in the Age of Ecology.* Cambridge, MA: The MIT Press.

Torres, Phil. 2016. *The End: What Science and Religion Tell Us about the Apocalypse.* Durham, NC: Pitchstone Publishing.

Zimmerman, Michael E. 1987. "Feminism, Deep Ecology, and Environmental Ethics." *Environmental Ethics* 9, no. 1 (Spring): 21–44. https://doi.org/10.5840/enviroethics19879112

2 Narrative Protocol and the Moral Turn

Qualitative Inquiry

This book is intended as an interjection, a prologue to a formative, catalytic conversation about the implicate universe we inhabit, about our place on Earth, and about the intractable dangers that threaten our tenure here. Several distinctive features of the ensuing presentation are notable. First, the 'unit of analysis' deployed here is the human population on Earth as a whole. The advantages of this approach are explained later in this chapter. From this holistic point of view, I have already introduced the premise that the material and transcendent dimensions of our shared experiences on Earth are not appropriately balanced.

Second, the argument undertaken in this book may be called a cultural critique. Admittedly, however, 'culture' is an evasive concept with no agreed-upon meaning in the social sciences or the humanities, making any critique of it somewhat problematic (Soini and Dessein 2016). In its most basic rendering, it may simply refer to another form of capital – the 'arts' – which add a fourth leg to the standard sustainability triumvirate of society, economy, and ecology. More broadly, culture may express a 'way of life,' the pragmatics of which serve the instrumental purpose of getting things done 'properly' according to tradition, mores, or style. On an even more comprehensive scale, it may refer to the shared process of semiosis – the quest for meaning – by which citizens of a given community understand and operationalize their deepest beliefs and values, a process which, significantly, will determine a community's preferred form of governance and its long-run evolutionary trajectory. This last interpretation, amenable most readily

DOI: 10.4324/9781032647067-3

to interdisciplinary and trans-disciplinary examination, is apposite to the present work.

Third, qualitative exposition is required because the self-induced peril we now face demands comprehensive, interpretive, even philosophical sensibilities to tackle and resolve:

> The knowledge at issue here is not the general, propositional kind. Rather, it is concerned with ... particular circumstances and dealing with all the complexity, ambiguity, emotions, and volitions entailed in these circumstances ... the kind of reasoning required here involves judgement, deliberation, and the assembly of a variety of empirical, ethical and political considerations necessary to cope with or make sense of the situation.
>
> (Schwandt 1997, xv–xxi)

Qualitative exposition is necessarily subjective and impressionistic and may, to the more analytic reader, lack objective substance and reliable empirical support. The more common positivist mode of knowledge acquisition, in its quest for universal or ahistorical laws of social science, seeks to discover quantifiable regularities in human activity, to develop theories that explain why those regularities hold under certain objective conditions, and to test those theories with reference to factual evidence. An interpretivist account of human behaviour, on the other hand, is concerned to explore and understand reasons for actions that derive from sets of interests which are shaped by characteristics of personal or social identity. It is a search for meaning, not data. I outline below the relevant features of qualitative research which lend clarity, coherence, and cogency to what Donald E. Polkinghorne has called "narrative knowing" (Polkinghorne 1988).

Like interests and identities, reasons for action are thoroughly embedded in social relations, often taking the form of shared beliefs which guide and legitimize behaviour. These beliefs or, more specifically, collectively held convictions and expectations, are deemed 'social facts' – facts with real causal power – but their unobservability, historical contingency, and evolution over time all conceal their presence from empirical inquiry. Nonetheless, they are important because they render human experience meaningful. "Narrative meaning," says Polkinghorne, "functions to give form to the understanding of a purpose to life and to join everyday actions and events into episodic units" (Ibid., 11).

The question of validation criteria is pertinent. Narrative forms of explanation may seem to be "arbitrary, subjective and soft" (Ibid., 94) compared to the deductive-nomological (law-seeking) explanatory scheme, but to be effective modes of epistemology and methodology must be commensurate with the ontological phenomenon at issue. Though its impact on the environment may be palpable, metanarrative itself is ontologically elusive; it is a belief system and set of ethical propensities, the existence of which is not possible to categorically affirm in a positivist moment of truth as correspondence. Alternative validation criteria of the interpretive mode must therefore include such elements as coherence and credibility; conformity with empirical evidence within the context of the story; the presence of key structural elements in the narration, such as agents, motives, and goals; and the progressiveness of meanings exposed, conclusions reached, or other contributions made to the inventory of human knowledge. Given these criteria, narrative is susceptible to an interpretive methodology that is sensitive to deeply implicit sociocultural conditions which can shape individual behaviour and legitimize public policy.

In Chapter 4 I will discuss the origin and impact of today's hegemonic metanarrative, the story that currently dominates the public consciousness. The taken-for-granted status of that story can be attributed in large part to its wholesale endorsement of growth and prosperity leavened with a disdain for any limitations that might be imposed by physical reality. This peculiar admixture of the material with an accommodating belief system that rejects material constraints allows the metanarrative to take on any shape or proportion and, I will show, also allows it to display the appealing but ultimately illusory appearance of a social, political, economic, and environmental panacea.

The Motive Force for Change

For better or worse, metanarrative provides the foundation of social normalcy and as such it is structural in nature. But, like the realm of meaning itself, it is also a dynamic system which is extrapolated by social forces through time and space, and which evolves through the constant and active interplay of new ideas.

But the role of ideas is often overlooked; instead, responses to the problem of agency are framed in pragmatic terms – how can individuals be educated and motivated, how can a popular movement be started and scaled up, what kind of infrastructure can support an upwelling of public concern and participation, and so forth. Prevailing wisdom suggests that citizen-led counter-hegemonic activism will, in due course, reach a critical mass which will overwhelm dysfunctional narratives and radically reform existing institutions, practices, and attitudes. This expectation, however, is countermanded by the reality that the business-as-usual paradigm still dominates the international agenda. Policymakers and captains of industry around the world are seized of the belief that no radical change is required, that adjustments at the margins within the dominant paradigm will suffice and that our current systems of governance are adequate to the task of creating a viable future. Normalized, legitimized, and institutionalized, the heavy forward momentum of modern industrial society is formidable.

Instead of taking a utilitarian approach, it may be that the driving energy for a truly transformative movement can be found in the creative humanities. People everywhere respond intuitively to stories, metaphors, and allegories that touch their lives, so morally sensitized emotional engagement presented in a literary (as opposed to an academic, technical, or pragmatic) format might open a more effective approach to the agency problem. After all, as William Ophuls reminds us

> We do not (and cannot) know what reality *is*. We can only know what it is *like* — that is, metaphorically ... [A]s a consequence of this utter dependence on metaphor, human beings are adapted to understand life not as formula but as story ... As obligate poets, we necessarily add feeling to understanding ... the human mind and heart crave myth and religion.
>
> (Ophuls 2011, 77–9, original italics)

With this in mind, then, a tectonic shift of the public mindscape towards the emotional, the metaphysical, and the moral might be the change we are looking for. A popularized dialectic which pits epiphany against epitaph, if it successfully penetrated the public consciousness, could have significant motive power.

Despite the imminent prospect of radical environmental disruption, ethical analysis in general, and environmental ethics in particular, have not prompted a tidal shift in globally held values, nor provoked the kind of massive change in human behaviour which many believe is an indispensable prerequisite for real planetary sustainability. Certainly, this speaks to the overwhelming magnitude of the problem at hand against which the humanities hardly seem a trenchant weapon for battle; but it also suggests that ethical debate might be too limited in its purview, too cautious in its consideration of alternative modes of attack.

One is reminded here of Kuhnian paradigm shifts which entail a transition from 'normal' to 'extraordinary' science. New modes of analysis are called for if familiar approaches become too hidebound, unable to cope with anomalies and novel problems presented by new information and experiences. At such pivotal moments in history, researchers sometimes turn to unorthodox methods and speculative hypotheses, making room for "a proliferation of competing articulations, the willingness to try anything, the expression of explicit discontent, and recourse to philosophy and debate over fundamentals" (Kuhn 1962, 91). Kuhn was mostly concerned with the natural sciences, but a related point of view also supports the perceived need at times of radical change to breach epistemological constraints. Normal versus extraordinary science is directly comparable in the social sciences and/or humanities with what Robert Cox calls "problem-solving versus critical theory." The former is conventional (solving a puzzle within a predetermined framework) but critical theory, on the other hand, stands apart from the prevailing order of the world and asks how that order came about, calling to question how and whether dominant belief systems might be in the process of changing (Cox 1986, 208).

The point here is that contemporary ethical discourse as it relates to the global environmental crisis, and indeed to global governance, is too self-consciously conservative. Despite the plethora of topical literature fraught with dire descriptive and prescriptive analyses, no cogent new insights with the power to penetrate the global veneer of normalcy have been successful at this task, and in this respect the discipline needs to sharpen its focus on the significance of the incipient radical disruption of the human enterprise on Earth, and on our moral standing as a species in that context.

Only having done so can the ideational component of governance – that is, the discursive content of the new metanarrative which we write for and about ourselves – lead the way to the better path we choose to follow to the future. Before taking on the question of our moral standing as a species, however, the following section will respond to the concern that the diversity of human society on Earth inevitably betokens a multiplicity of narratives, not just one.

One or Many

By using inclusive words such as 'we' and 'our' in most of the foregoing text I have referred to our species on Earth as a singular entity. This may seem inappropriate, or simply wrong. It is quite clear, after all, that multiple civilizations and communities around the world exhibit unique histories, cultures and varieties of economic and political systems which present in their totality a kaleidoscopic image of human society, the diversity of which is one of its most important features. And it seems equally clear that multiple narratives about life on Earth co-exist right now, narratives about money and power, about democracy and human rights, about feminism, environmentalism, the information revolution, and of course about religion and various preferred ways of being and personal living. To overlook this rich diversity is to do a disservice to the manifold complexities of human inventiveness and adaptive resilience.

On the other hand, however, highlighting diversity can obscure the general form, function, and character of the whole which comprises those various parts; one can lose the forest for the trees, to employ a familiar aphorism. The diversity lens brings to focus a medley of human social projects, and along with that a complicated matrix of dynamic relationships among those projects. By gathering all these elements together the whole point of view brings to light features which are unique to the aggregated totality, such as the fact that one incumbent story – one *meta*-narrative – covertly rules the contemporary mindscape; and that our species in its entirety is embedded in a planet-wide socio-ecological system which may or may not evolve according to our wishes and requirements. On this account, it is useful, even necessary, to posit a singular entity called 'human society' which now finds itself entangled in a self-induced existential struggle on one indivisible planet.

In principle, of course, we can enjoy the benefits of both unity *and* diversity. There is no reason to suppose that communitarian values cannot be respected within a cosmopolitan framework, no reason why differences cannot flourish within a shared ethic of sufficiency. There is no reason why micro-variability cannot be conjoined with macro-stability expressed in a new metanarrative about the long-term relationship between Earth and us. No reason in principle, that is, but in practice the balance between unity and diversity, between parts and the whole, is very hard to find and to maintain. For example, those who prize above all else individual freedom may be forcibly constrained by an invisible social structure which privileges community solidarity, shared values, and the pursuit of common goals. The resultant tension between individual freedom of choice and public responsibility sets up a problematic relationship between 'the one and the many,' the resolution of which has challenged philosophers for millennia.

Interestingly, no such tension exists in the material world of nature. Individual plants and animals live in nested webs of ever larger and more complex ecosystems, a hierarchy of wholes which simultaneously enables and constrains opportunities to thrive by efficiently transferring information and energy among component parts. The vitality of the total global ecosystem is enlivened by the constant interplay of competition and cooperation among those parts, and the ensuing dynamic balance forged by nature (called mutualism) sets the stage for individual lives and species to flourish as best they can. The success of ecosystems to strike a balance between diversity and unity has inspired the hope that human social systems might be able to imitate that success by utilizing the principle of 'biomimicry.'[1] Reminiscent of Aldo Leopold's land ethic (Leopold 1949) and Jim Cheney's discussion of bioregionalism (Cheney 1989), the aspiration to imitate nature is entirely sensible and could lead to the discovery of important lessons for human survival.

Hopeful though this may be, however, it ignores the fact that nested social structures are products of the human imagination not necessarily beholden to nature's template. For us, the embedding of difference in a functional whole is a matter of choice; we may or may not choose to do it. And, moreover, because the character of interlaced social systems is informed if not wholly determined by the dominant stories of the day, we may or may not do it well. Stories which tout individual prowess, racial purity, or disdain for the 'other,' for example, may not be suited to the successful accommodation of difference. It is important to acknowledge that socially constructed narratives, especially metanarratives, are not intrinsically or necessarily optimal or, for that matter, benign.

In fact, examples of positive change instigated by a clash of foundational ideas are rare in history. To cite a recent example, the Cold War between the United States and the Soviet Union in the wake of the Second World War was a contest of strength, but it was also an argument about the relative merits of capitalist and communist socio-economic systems. Each was advertised to better serve the common good, and each was offered to the world as the normative core around which global governance should evolve. Capitalist ideology won that argument to Western applause but to the dismay of many modern critics who now see capitalism as rapaciously amoral and a root cause of environmental degradation.

The Protestant Reformation offers another cautionary tale of transformative change in history. In 1517 Martin Luther tacked his 95 Theses to the door of Castle Church in Wittenberg, Germany, launching a frontal assault on certain practices (especially indulgences) of the Roman Catholic Church. With the help of the printing press– a broadcast innovation not unlike social media in today's world – the Protestant movement was a game-changer. The content of the theses was of course not germane to planetary sustainability, but nonetheless the effect of this assault was revolutionary. That the power of the Church and Catholic dogma during the ascendancy of the Holy Roman Empire could be challenged, if not toppled, by an opposing set of beliefs speaks clearly to the kind of ideational battle for which today's new global narrative must be fully prepared.

As noted, however, the clash of contrary ideas, especially those contending for metanarrative status, is inevitably jarring, even dangerous. Recall that the Reformation stoked violent religious conflict across Europe, culminating in the Thirty Years' War. Recall also that Luther's failure to convert European Jews to Christianity (the failure of one narrative to subsume another) prompted him to adopt a most appalling kind of anti-Semitism which, centuries later, was called upon to inform and legitimize the National Socialist movement in Nazi Germany (Probst 2012). This is all to say that words, beliefs, shared commitments, and new stories are tremendously powerful phenomena, and dominant stories always carry the risk of oppression. One need only think of *lebensraum*, the innumerable atrocities committed in the name of religion, or, looking forward, the possible rise of global environmental eco-fascism, a forcible claim to the right to mete out scarce resources in the name of sustainability. To avoid these outcomes, the new metanarrative many of us anticipate must be founded on the most secure moral footing. This aspiration, however, has not yet been realized.

The Moral Turn

Growing public concern hints at the worry that what we have done to the planet may not only be negligent and misguided, but wrong. If that is true, any self-induced, life-damaging eco-catastrophe (such as runaway climate change) would constitute an explicitly immoral outcome of human behaviour. And because immoral outcomes are necessarily preceded by immoral causes, this supposition implies that an accurate diagnosis of anthropogenic planetary ecological instability should originate from the critical scrutiny of our current behaviour from a normative point of view, and with a particular focus on the metanarrative which legitimizes it.

Societies around the world have long been sullied by morally dubious behaviour. War, greed, radical economic disparities and distributional injustice, patriarchy and the oppression of women, child exploitation, cultural persecution, and genocide, among many other examples, are all deeply disturbing concerns. The fact that we can perceive the immorality of these behaviours – that we recognize them as aberrations – indicates that we have access to a variety of frames of reference which anchor our sense of value, and with regard to which we can make moral judgements. Applying such judgements to real-world conflicts sets up the inevitable tension between right and wrong behaviour, between what is and what should be. This tension is a constant feature of the human condition, driving moral discourse since time immemorial.

Recurring ethical quandaries test our mettle, keeping us alert and morally engaged, and in that sense they are beneficial. But some quandaries extend beyond the inconvenient, unfair, or tragic to the existentially lethal – and that is what takes today's moral discourse to a new level. Two such examples of lethal danger are particularly egregious: the destruction of other species and their habitat, and the foreclosure of our own future.

We depend for our lives on flora and fauna, just as they depend on each other. To diminish biodiversity is to diminish the life chances of all. But more to the point, when we humans arbitrarily terminate other life forms, we arrogate to ourselves undeserved power. Mature exercise of that power might include the final elimination of, for example, pathogenic viruses, but no such maturity is visible in the wake of the Sixth Great Extinction, now underway, of which we are the thoughtless perpetrators (Kolbert 2014). This destructive behaviour insults life, weakens Earth's vitality, erodes sustainability, and is emphatically unconscionable.

But our anthropocentric zeitgeist and inflated sense of exceptionalism render the planet-wide loss of biodiversity almost invisible. These features also blind us to the possibility of our own self-destruction instigated by a ruinous, cascading eco-catastrophe. Such an event would nullify any concept of sustainability. It would entail the end of human possibilities, the end of the exploration of the human experience, the foreclosure of our own future. There is nothing to be gained here by dwelling on the brutalities, including climate change, we have committed against nature, and against ourselves, which foreshadow the potential extinction of our species but this much can be said: if modern industrial society marks the last of the great civilizations which we have produced on Earth over the last 5,000 years or so, then our remarkable success as a species will have become a demeaning failure of will, of merit, of foresight, and of moral intelligence. This too would be emphatically unconscionable.

Recurring cases of morally problematic behaviour are, as indicated above, instructive from a normative point of view, and some do have serious environmental consequences. War despoils the landscape, greed causes excessive consumption, patriarchy diminishes the feminine impulse to nurture, and impoverishment leads to the unsustainable harvesting of scarce resources; but, over the millennia, these examples of harmful behaviour have not raised the spectre of extinction. The expropriation of the lives and living spaces of other species for the benefit of the human animal, however, and the expropriation of our own future for the embellishment of the present are another matter. The wanton razing of life on Earth combined with (even the possibility of) the terminal interruption of the flow of human history should now be the main ingredients of contemporary moral discourse. To this end, a new frame of reference is required.

Many good people agree that a transformative move to planetary sustainability is still possible. They declare that the world is in trouble but sustainability is within reach if urgent and concerted international action is taken immediately. They wait hopefully for an effective pre-emptive response from political leaders. But policymakers around the world have been unmoved by calls for radical change. Instead they have been marshalling their collective will in pursuit of objectives defined for them by the dominant story of our time. By all means available, they wilfully pursue the business of Progress and Prosperity. This will not change, nor will a calamitous result be avoided, until that story is subsumed within a larger, better, and morally enlightened context. Until the new global metanarrative gains clarity, cogency, legitimacy, and popularity, the old story will continue to dominate our mindscape, leading us quickly down the wrong path.

Note

1 Coined by Janine Benyus. See biomimicry.org for details.

References

Cheney, Jim. 1989. "Postmodern Environmental Ethics: Ethics as Bioregional Narrative." *Environmental Ethics* 11, no. 2 (Summer): 117–34.

Cox, Robert W. 1986. "Social Forces, States and World Orders: Beyond International Relations Theory." In *Neorealism and Its Critics*, edited by Robert O. Keohane, 204–54. New York: Columbia University Press.

Kolbert, Elizabeth. 2014. *The Sixth Extinction: An Unnatural History.* New York: Henry Holt and Company.

Kuhn, Thomas S. [1962] 1996. *The Structure of Scientific Revolutions.* Chicago, IL: University of Chicago Press.

Leopold, Aldo. 1949. *A Sand County Almanac.* New York: Oxford University Press.

Ophuls, William. 2011. *Plato's Revenge: Politics in the Age of Ecology.* Cambridge: The MIT Press.

Polkinghorne, Donald E. 1988. *Narrative Knowing and the Human Sciences*. Albany, NY: State University of New York Press.

Probst, Christopher J. 2012. *Demonizing the Jews: Luther and the Protestant Church in Nazi Germany*. Bloomington, IN: Indiana University Press.

Schwandt, Thomas A. 1997. *Qualitative Inquiry: A Dictionary of Terms*. London: Sage.

Soini, Katrina, and Joost Dessein. 2016. "Culture-Sustainability Relation: Towards a Conceptual Framework." *Sustainability*, 8, no. 2: 167. https://doi.org/10.3390/su8020167

3 Global Governance in Contemporary Culture

Metanarrative and governance are closely related but not identical concepts. I have described the former as the command-and-control function of modern human society which informs the choices we make and the aspirations we share. Socially constructed beliefs and values contribute to our sense of meaning and purpose, and thereby to the development trajectory we choose to authorize, and more broadly to the character of the human/nature complex we inhabit, which is to say, the kind of relationship we have with our terrestrial home. Governance, on the other hand, includes both a practical and an ideational component and the one is often mistakenly conflated with the other. Metanarrative refers to the latter. The practical component of governance is that array of rules, practices, methods, and tools which operationalize and implement the precepts stipulated by the metanarrative, that is, by the collective intentions and ethical propensities inherent to the ideational component of the system of governance. Given a certain sense of urgency regarding the instability of our relationship with Earth and the apparent need to chart a new path forward to long-run planetary sustainability, it is appropriate now to re-evaluate, and reformulate as necessary, the social constructions that guide our practical and political behaviour.

This approach to better governance is intended to contextualize and catalyse the myriad hands-on activities aimed at securing the long-run material sustainability of the human/nature complex such as moving to a low-carbon energy regime, building resilient communities, alleviating poverty, and so forth. These projects are ongoing and clearly necessary but they are

DOI: 10.4324/9781032647067-4

also slow-moving, standing in sharp contrast to repeated (and increasingly strident) calls for radical, transformative change in our relationship with Earth, and in the direction of our forward motion. The proposition on offer here is that this radical change, tantamount to securing the safety and the future of human life on Earth, will arise in large measure from a revitalized conception of global governance with a fresh focus on its ideational component. This shift in focus from the material to the ideational, from the physical to the metaphysical, does not diminish the need for effective government but it does move to the forefront the broader objective of understanding more clearly the existential implications of our current situation. Positions adopted regarding these implications (whether thoughtful or dismissive) give meaning to the human condition, and underlie all modern modalities of governance.

Global Governance: What It Is, and What It Does

Notwithstanding the territoriality of sovereign states and other claims to private property, Earth's biosphere is a public space. Its protection is a public and therefore a political responsibility. Managing the relationship between people and planet is the centrepiece of that responsibility and, from a whole-Earth point of view, the defining challenge of global governance. The term 'governance,' however, enjoys no concise and common definition, and the same is certainly true for the larger and more comprehensive concept of 'global governance.' In fact, as a first approximation, if governance were defined straightforwardly as the service that government delivers, then absent a world government global governance would not exist at all. And yet, a broad and diverse literature attests to its theoretical existence, and the relatively high degree of order which permeates and stabilizes world affairs speaks to its empirical presence as well.

By some accounts, global governance is about the exercise of authority on the world stage. Those who wield legitimate power include the formal institutions of national governments, a variety of international organizations (including the United Nations), a plethora of non-governmental organizations, and a dense web of formal and informal networks including sub-national, transnational, and supranational organizations that increasingly contribute to the establishment and functioning of global rules, norms, and practices. This broad definition describes a multi-layered, multi-purpose system of governance, the ultimate objective of which is to manage a variety of global problems. Given this diffuse form of governance, however, plus the fact that the global polity is fractured into more than 200 sovereign territorial states, each commited to exercise its own brand of self-determination, that management is bound to be inefficient. As John Ruggie wryly remarks, "[I]nternational management [especially regarding environmental challenges] is likely to be supplied in sub-optimal quantities even when all concerned agree that more is necessary" (Ruggie 1998, 144).

Another point of view focuses less on actors and more on what policymakers believe to be true, arguing that global governance implies a single, coherent worldview and the (more or less) universal participation of all states in an agreed programme of action. For example, contemporary global governance may simply refer to the ascendancy of a neoliberal ideology which grants priority to the global economy, lionizing the efficiency and effectiveness of market outcomes and informing the operational behaviour of policymakers and civil societies everywhere. Neoliberalism is the common currency of today's economic relations and in that respect it certainly is a form of governance, but its narrow focus on utility suggests that its dogmata are not well-matched with a fuller conception of public purpose.

Adopting a more aspirational point of view, there are those who believe that the essence and the most desirable expression of governance can be found in the participatory impulse advocated by liberal democratic governments and related institutions and organizations around the world. On this view, global governance is about, or should be about, people power, entailing the personal involvement of, essentially, every one of us. We live in a globalized world, so the argument goes, trending to a form of direct democracy built upon the informed and responsible participation of 'global citizens.'[1] The most important actor in governance systems is the individual, not organizations, and the aspirational destination is a democratic planetary civilization, including a parliamentary-style United Nations,[2] living in harmonious relationship with the ecosphere. I take a closer look at this conjecture in the latter part of this chapter under the banner of postmodernism.

Digging deeper into the academic literature pertaining to global governance would take us too far afield so, for present purposes – that is, for a discussion about the relationship between people and planet – it will be easier to talk about what global governance does, as opposed to what it is. Here the field is simplified dramatically.

Global governance, regardless of how it may be constituted, can be thought of as serving two key functions. The first of these is the provision of public order for the public good in the international arena. Business and finance, trade and transportation, health and safety, human rights, the environment, arms control, and cyber-security are just a few global issue areas which are regulated, managed, or otherwise overseen by regimes intent on providing necessary services and administrative predictability for all those affected. The second function of global governance, and the one central to this book, is to provide a generalized steering capability for human society. Notwithstanding its many interpretations, the verb 'to govern' is originally drawn from the Greek *kubernān* and the Latin *gubernāre*, both meaning 'to steer.' Steering implies two things: plotting a safe course through the obstacles of everyday life and, equally or perhaps even more importantly, moving steadily toward a chosen destination.

Where Are We Going?

Good governance is intentional. To the extent that world affairs are guided by systems of governance, they are shaped fundamentally by a collective sense of what purposes should be pursued, and what kinds of values should prevail. Global governance, in other words, provides order for a reason, namely, to achieve an agreed objective. An agreed objective will necessarily conform to a given set of values (one objective is, by some standard of valuation, better than another), so the purposeful nature of governance unavoidably includes a normative dimension. "Norms are at the heart of all governance structures," as one analyst has succinctly expressed this important point (Bernstein 2001, 5).

Prosperity defines our preferred objective and progress identifies the means to attain it, but our rapid forward motion to a better future, though well-meaning, has been problematic. Human behaviour relative to Earth's ecosphere and resource base is demonstrably (and exponentially) rapacious. Population growth and rampant poverty still run far ahead of material well-being and many thousands of our fellow species on Earth have already been lost, their habitat destroyed. These unintended consequences are disturbing and counterproductive, calling to question the propriety of the pursuit of progress and prosperity. Though emblematic of modern global governance, this programme for human advancement is evidently not taking us where we need to go.

Any new metanarrative about planetary sustainability must emerge from an expansive context, one which in fact encompasses the whole Earth. Today that context is the Anthropocene epoch, the self-proclaimed apex of our hegemony on Earth and in fact a new geological interval which intimately fuses two major components – Earth itself and the human population which thrives upon it. But how are we to make sense of the onset of the Anthropocene, and what kind of story does it tell? Are we masters of our own fate, headed to a secure and productive relationship with the planet, or are we a rogue species rushing mindlessly towards eco-catastrophe? The human capacity for foresight and agency in combination with the logic of self- and system-preservation suggests a hopeful future. Physical and ecological evidence to date supports a less sanguine conclusion. Which of these points of view is closer to the truth? Do we shape our future by considered choice or have we been victimized by an innate propensity to exploit, maximize, and manipulate regardless of the consequences?

A heuristic device is available to help assess these two very different perspectives (Hollis and Smith 1990). The device distinguishes between explaining the Anthropocene from an outside or objective point of view, and understanding it from an inside or subjective one. The former captures the biophysical (genetic, ecological) factors which contribute to the evolution of the relationship between people and planet. The latter captures the sociocultural factors (choice, intention) which inform and guide that process. Together they suggest two kinds of story about the direction of our forward motion, invoking the familiar distinction between positivist and interpretivist modes of discovery; the relationship between them is of particular interest.

Explaining the Anthropocene

The outside or objective point of view sees humanity – the set of all human beings – in much the same way as an ecologist would see any other species living on Earth. Ecologists study observable behaviour only, making testable inferences that explain the patterns they see. No allowance is made for intention or meaning. Ecology is a thoroughly empirical science. On this view, the ascent of the human population has followed a predictable, exponential trajectory.

Any and all species will proliferate within the bounds of their respective niches. Niche boundary conditions are defined in terms of external and internal constraints. External constraints are imposed by geography, climate, food and water supply, competition, predation, and so on. Internal constraints are determined, essentially, by the genetic complexity and morphology of the species in question. Neither of those limits is problematic for us. We possess a unique and very powerful set of adaptive capabilities, and a sophisticated intelligence which obliterates boundaries. We can live anywhere, eat anything, and take whatever we want. Our rise to the top of the world is, from an ecological point of view, not at all surprising; in fact, it is entirely predictable. We are a super-species, and that is the natural cause of our domination of the planetary landscape. Simply put, we grow and flourish as a species because we can.

A surging population, the appropriation of habitat and the extirpation of other life forms, the despoliation of land and water, even the massive disruption of planetary operating systems are, however, the unhappy consequences of our exceptional endowments. Explaining the onset of the Anthropocene from the outside point of view conjures an unflattering image of an invasive species overwhelming local ecosystems and food supplies, radically altering the landscape, collapsing affected populations, and, ultimately, debilitating the invader too. It exposes a dynamic trajectory driven by natural forces which operate beneath the threshold of perception but which are nonetheless deeply embedded in the patterns of life on Earth. It suggests that we are being led by a biological imperative, not considered purpose.

Understanding the Anthropocene

In contrast to the outside/explanatory mode, the inside point of view makes full allowance for socially constructed features of human society on Earth such as rules of behaviour, intention, and meaning. Here we seek not to explain objectively, but to understand implicitly. Instead of looking for causes of behaviour, we look for reasons. On this view, the human enterprise is a malleable social project amenable to effective management based on choice. The subjective point of view brings internal psychological and sociocultural factors into view, the qualities that define us as creatures fully capable of constructing our own socio-ecological relations. This approach suggests that natural impulses can be overridden, purposely managed in the interests of survival and for the benefit of the full realization of our potential as a transcendent species. It implies that the Anthropocene epoch might open exciting new opportunities for us to rewrite the story of the evolution of human society on Earth.[3] And it promises that collective human agency should, can, and will restrain and redirect our exuberant but heretofore destructive propensity to flourish, emphatically exposed by the outside perspective.

An ethic of sufficiency internalized by each of us is a necessary component of sustainability but, though an important part of the lifestyle of scattered individuals and communities around the world, no such ethic can be attributed to human society as a whole. In fact, the opposite is true. Instead of reining in expansionism, human agency today is fully aligned with and adds tremendous impetus to this atavistic impulse. The outside view stipulates that we grow and flourish as a species because we can. Now, another dynamic comes into play, this time from inside human society: we grow and flourish because we want to.

This is evident in the purposeful pursuit of material prosperity. The manifestation of a uniform belief system shared right around the world, this quest stipulates that growth and abundance serve important social ends. Different states and cultures may operationalize it differently, but in our collective political and economic lives nothing commands our sustained allegiance like the pursuit of growth, progress, and prosperity. While the natural (external) dynamic aggressively drives the assimilation of habitat and resources, the social (internal) dynamic adds desirability and legitimacy to the process of relentless expansionism. These two motive forces are congruent and mutually reinforcing. They work in tandem, they are both deeply implicated in the onset of the Anthropocene, and together they make clear why a new modality of governance informed by a better metanarrative has become essential.

Narration in Modern Culture

The green shoots of new stories are already abundant,[4] springing up around the world. But none, individually or collectively, has seriously challenged the incumbent story of Progress and Prosperity, or subsumed it. The new global narrative has not been written yet. David Korten has offered one impression of what it might look like. Here is what he says:

> Whether specific details of our chosen story are right or wrong is less important than whether its overarching narrative awakens us spiritually; inspires cooperative, mutually beneficial relationships; supports a way of living that recognizes the wonder, beauty, goodness, ultimate meaning and value of life; and puts us on a path to a viable future. Most important at this moment in the human experience is that our chosen story calls us to accept adult responsibility for the consequences of our choices for ourselves, one another, and a living Earth.
>
> (Korten 2013)

This is a good start, and heartfelt no doubt. Most green shoot stories rising around the world share similar sentiments. They offer clear, ethically informed visions of the future and prescribe new collective projects or development models for the world. They are all suffused with hopeful expectations, with an optimistic outlook, and with calls for responsible change. But the new global narrative will not be a simple story portending happy endings, nor will it promise an easy transition to planetary sustainability. If that were the case, then the multitude of such stories now in circulation would surely have taken us further down the road towards the goal of stabilizing the still-precarious relationship between people and planet. To realize and embrace the sentiments named by Korten is necessary, but it is by no means sufficient. The new story, traveling beyond those sentiments, may very well be darker, tougher, evoking deeper and more vivid currents of thought and feeling, confronting directly the hard truth that we are a species in existential peril.

The Postmodern Malaise

I mentioned earlier an aspirational form of global governance which highlights the story of the individual as 'global citizen.' This is laudable (from a liberal democratic point of view) and important insofar as it speaks to the independent power of the individual as an agent for change. I will take up this theme later in my discussion of agency, noting in the meantime that there is today considerable tension between individuality and the hegemonic metanarrative which currently informs our system of governance. This tension takes the form of a postmodern malaise which explicitly rejects any kind of universalizing perspective.

Jean-François Lyotard popularized the term 'postmodernism' in his seminal book published in 1979. "I define *postmodernism*," he said, "as incredulity toward metanarratives" (Lyotard [1979] 1984, xxiv, original italics). Primarily a product of Western civilization but representative of a much wider constituency, Lyotard argued that the postmodern condition speaks to a loss of faith in modernity – that is, to a rejection of the metanarrative of progress, broadly defined in terms of the ascent of Enlightenment rationality, the promised emancipation of the world's exploited, and the creation of wealth. Instead of these felicitous outcomes, we have seen two world wars, the rise of a transnational elite which controls the levers of political and economic power, radical economic disparities separating North and South, the increasingly heavy influence of technology over everyday lives and, not least, the spectre of a global environmental emergency. For postmodernists, the lure of 'prosperity for all' is vacuous, a nominal sheen on a false and misleading story. By rejecting it they bluntly reject any metanarrative which purports to identify a common goal, a transcendent truth, or a universalizing perspective.

The grand narrative of progress has not been dismissed across the board, of course (Pinker 2018). It is still promulgated by those who tout its benefits, especially political leaders whose electoral success depends upon it, by those who aspire to wealth and power, those who enjoy conspicuous consumption on a massive scale, and also by ordinary people who yearn hopefully for a better, more affluent lifestyle and a prosperous future ahead. Despite this widespread support, however, the programmatic pursuit of progress and prosperity is vulnerable not only to the broad-brush criticisms levied above, but also to a cultural critique which paints modernity as a tumultuous era characterized by speed, by the onrush of new events, by the flickering ephemera of television, by computers and smart phones, and by the lionization of personal accomplishments, all of which create a simulated, surrogate reality and a discontinuous social experience. Continuity with the past is lost, the sense of geographic place is compromised by virtual travel, the future is unpredictable, and the only timeless law is the disconcerting notion that only change is constant.

The information revolution has played no small part in this upheaval. Richard Kearney makes this important observation regarding the impact of the information revolution on cultural continuity:

> The shared *experience* of traditional communities, based upon the transmission of inherited stories, myths, legends, tales, lore, from generation to generation, is being replaced ... by the more anonymous and instantaneous transmission of *information*. This signals the end of cultural memory with its specific qualities of continuity, authenticity, accumulated depth and wisdom, and the beginning of an electronically interconnected network — the so-called global village.
>
> (Kearney 1997, 184, original italics)

Access to information and the attendant tools of dissemination is undoubtedly important and maybe, as Kearney goes on to say, the information revolution contains "the seeds of a new universalism transcending the tribal divisions and differences of traditional national cultures" (Ibid., 183). In the meantime, however, we are still living in "a society of spectacle and pseudo-events, a society in which we seem to be having difficulty distinguishing between narratives which are genuine or fake, enabling or disabling, better or worse" (Ibid., 184). These problems are exacerbated by anonymity on the internet which has freed actors from social constraints on behaviour and expression, and by chat rooms that facilitate the fomentation of misanthropic or subversive beliefs by otherwise isolated individuals.

Postmodernism rejects but is nonetheless a product of modernity. Indeed, it is ironic that the shattering of time and space has been assimilated by postmodernists even as the grand narrative of progress itself has been rejected as dysfunctional. Difference and individuality are celebrated and fragments of life are re-purposed as nodes of personal proprietorship; each of us becomes the locus of our own experience and the creator of our own reality. Optimistically, Arran Gare sees in this the possibility of a postmodern cosmology which, drawing on Alfred North Whitehead's process philosophy, privileges becoming over being (Gare 1995). What this means is that postmodernists should (and do) avoid the reification of abstract entities – no absolute truths or principles, no Platonic forms, no teleological tendency to moral evolution or maturation – in favour of a personal process of becoming understood as an exercise in immanent- or self-creativity. In this way, "life itself is lived as an inchoate narrative" (Ibid., 137) and from the interplay of diverse stories a multi-dimensional narrative might arise in which individuals "experience themselves as processes of becoming actively participating in the becoming of the world" (Ibid., 155). This is actually an attractive cosmology which anticipates the concluding chapter of this book, with the important difference that postmodernists struggle to find virtue in an atomized world, whereas I will suggest that personal agency can benefit from (and partake in) the 'reification of abstract entities' mentioned above, a process which offers a more useful and morally sound platform for (r)evolutionary change.

Notwithstanding this friendly reading by Gare, Foucauldian post-structuralism (a leading intellectual voice of postmodernism) insists that narratives of any kind, especially metanarratives, are really a form of false consciousness and consensus, universalizing tendencies the covert objective of which is to suppress, not encourage, difference and individuality (Foucault 1984). On this view, history proceeds from domination to domination, each episode driven by competition for the hearts and minds of people. At pivotal junctures when discourses clash, the contest is resolved to the degree that coercion is effective and consent is temporarily achieved, and new modes of normalization are imposed. This recurring pattern of social evolution can and should be opposed according to post-structuralist doctrine. In the modern context the result of opposition to the grand narrative of progress has been the assimilation of social fragmentation and the celebration of it as a core feature of an individualistic lifestyle.

Because this interpretation tends to supersede Gare's more forward-looking opinion which foresees the benefits of a 'multi-dimensional narrative,' postmodernism as a cultural phenomenon is commonly regarded as self-consciously reclusive, tending not to unite people but divide them. In that sense it is not a powerful political force, nor does it offer a coherent methodology for understanding how to achieve any kind of consensus regarding the environmental crisis. It is, in short, antithetical to global governance.

Paradoxically, another major current of habitual behaviour in modern culture which militates against narration pulls in the opposite direction. Unlike postmodernism, positivism is a prevalent mode of discovery around the world today and a powerful political force. It relies upon sensory experience and embraces rationality as an instrument to gain knowledge and to achieve control over the natural world for the benefit of humanity. It seeks objective and timeless laws via observation, measurement, and a quantitative methodology. Inasmuch as human affairs are part of the natural order, these too are amenable to positivist inquiry without any needful reference to normative (immaterial) considerations. Because this approach is widely believed to be progressive and, ultimately, capable of solving most worldly problems effectively and efficiently, it neatly combines nomological determinism with an optimistic view of the future.

Positivism is a product of Enlightenment thinking, developed canonically by Auguste Compte in the early 19th century; it provides the methodological and ontological foundations upon which most science today is taught and practised, and its rewards have been obvious and plentiful. It has, on the other hand, been roundly criticized for rejecting as irrelevant anything unseen and immaterial, a rejection which has baffled sociologists and critical theorists alike. The Frankfurt School,[5] for example, has seriously challenged the notion of rationality, concluding that most humans are anything but, and has disparaged the entire movement of empirical positivism as an ideology, as 'scientism' which inappropriately privileges a materialist belief system over breadth of understanding and common sense. Critics also warn of the loss of aesthetics and morality, and decry the 'disenchantment' of nature which makes Earth vulnerable to the instrumental aggression of practical reason. By relying solely on empirical data and rejecting as fanciful metaphysics, theism, introspection and intuition, critical theorists in general argue that positivism creates a world in which narrative has no role to play.

These contrary trends – postmodernism and empiricism – both stand opposed to the evolution of a new story, the first touting local fulfilment, the other pointing to the material world as the only reliable measure of utility; yet both have something useful and important to contribute to the challenge of promoting a historically pivotal paradigm shift, namely, their shared focus on the individual person as the ultimate locus of agency. I will develop this theme in Part II of this book but, because it is relevant to the composition and character of political leadership as it pertains to narration, I offer the following introductory comments.

Democratically elected individuals are charged with the responsibility to reflect and enact the wishes of the electorate, to bring to fruition their common objectives, and in this sense successful representation only requires certain essential functional traits such as honesty and competence. If a popular new sentiment bubbles up from below, then putting its precepts into action is just a matter of accommodating the wishes of the people. This fits well with the standard (positivist) conception of political leadership as law-abiding, evidence-based, and practical.

This simple version of democracy, however, can run into serious trouble if, for example, the hypothetical 'new sentiment' just mentioned had been popularized by a charismatic demagogue whose intentions were self-serving or malevolent. In that case, one should reasonably expect elected representatives (or any other kind of responsible leader for that matter) to recognize non-authentic public pressure and to respond accordingly; and here a contrary feature of leadership becomes apparent, namely, the need to act independently to preserve the public good. Leaders may, in their role as individuals whose judgement is respected, see fit to rebuke public expectations if and when, in their opinion, the electorate has been misinformed or misled, or if the population at large has been swept up in a rash and irrational moment which may, in the longer view, cause public harm.

As high-minded as this may appear to be, contemporary emphasis on individual autonomy and a condescending treatment of the ideational domain affords undue power to politicians by failing to explicate and secure the moral grounding for such 'supra-democratic' independent action. Absent a sound and living social ethos, leaders may be corrupted by private interests or personal ambitions, or their capacity for coherent evaluation may become dissolute and apathetic, so the freedom to act independently of the electorate will be dangerous, not useful. In such cases, instead of acting upon clearly expressed and socially cogent argument, leaders will deceitfully declaim that they are acting 'on principle' or that they are 'not trying to win a popularity contest' to justify their actions. And this in turn inevitably raises that perennial question so important to the long-run maintenance of social integrity: who will watch the watchers? And what exactly lies at the bottom of this bottomless regression?

The postmodern brand of individuality does not fare much better. Arran Gare's optimistic expectation of diverse stories in an atomized world congealing into a multi-dimensional narrative (as discussed earlier) cannot happen coherently, in my view, without the 'reification of abstract entities,' in other words, without a sound footing and a clear understanding of the source and substance of moral behaviour. The acknowledgement (shared with positivists) that individual human lives are the necessary locus of agency is important – indeed, critical for planetary sustainability – but the modern propensity to isolate and lionize humanity, as individuals and as a species, has fostered a sense of exceptionalism that goes far beyond what is earned, or deserved. Vaclav Havel offered a useful corrective for this when he reminded us that "We are not here alone, nor for ourselves alone."[6]

Part II of this book responds to this challenge by embracing all the 'soft' and, perhaps, idiosyncratic features of qualitative inquiry, and by looking for the moral foundations upon which social stability, good governance, and effective agency must ultimately rest. Before doing so, however, the next two chapters will bring the incumbent metanarrative into sharper relief, exposing its most vulnerable attributes.

Notes

1 Education is an important requirement for the evolution of global citizenry, as recognized by a growing movement called 'ecopedagogy.' For more on this, see Misiaszek (2020).

2 See the World Federalist Movement at: www.wfm-igp.org

3 Ecomodernists certainly think so. For proponents of this view, the Anthropocene is not a crisis but the beginning of a great new era of human-directed advancement and agency. See www.ecomodernism. org/manifesto

4 David Korten (2013) and Thomas Berry (2003) have made notable contributions to this genre. See also www.newstories.org; www.findh orn.org/programmes/living-the-new-story/; https://journeyoftheunive rse.org/; www.metanexus.net/ritualizing-big-history

5 For a brief introduction, see www.britannica.com/topic/Frankfurt-School

6 Vaclav Havel, Acceptance Speech for Liberty Medal, Philadelphia, USA, July 4, 1994. Available at https://constitutioncenter.org/libertyme dal/recipient_1994_speech.html

References

Bernstein, Steven. 2001. *The Compromise of Liberal Environmentalism.* New York: Columbia University Press.

Berry, Thomas. 2003. "The New Story." In *Teilhard in the 21st Century: The Emerging Spirit of Earth*, edited by Arthur Fabel and Donald St. John, 77–88. Maryknoll, NY: Orbis Books.

Foucault, Michel. 1984. "Nietzsche, Genealogy, History." In *The Foucault Reader*, edited by Paul Rabinow, 76–100. New York: Pantheon Books.

Gare, Arran E. 1995. *Postmodernism and the Environmental Crisis.* New York: Routledge.

Hollis, Martin, and Steve Smith. 1990. *Explaining and Understanding International Relations.* Oxford: Clarendon.

Kearney, Richard. 1997. "The Crisis of Narration in Contemporary Culture." *Metaphilosophy* 28, no. 3 (July): 183–95.

Korten, David. 2013. "Religion, Science, Spirit: A Sacred Story of Our Time." *Yes! Magazine*, January 18, 2013. www.yesmagazine.org/happin ess/religion-science-and-spirit-a-sacred-story-for-our-time

Lyotard, Jean-François. [1979] 1984. *The Postmodern Condition: A Report on Knowledge.* Translated by Geoff Bennington and Brian Massumi. Minneapolis, MN: University of Minnesota Press.

Misiaszek, Greg Williams. 2020. *Ecopedagogy: Critical Environmental Teaching for Planetary Justice and global Sustainable Development.* London: Bloomsbury Academic.

Pinker, Steven. 2018. *Enlightenment Now: The Case for Reason, Science, Humanism, and Progress.* New York: Penguin Books.

Ruggie, John G. 1998. "Introduction." In *Constructing the World Polity: Essays on International Institutionalization*, edited by John Gerrard Ruggie, 1–44. London: Routledge.

4   The Story of Progress and Prosperity

Ethnographers refer to formative discourses or metanarratives as cosmologies by which they mean descriptive accounts of how and why things are the way they are within a given culture. Often taking the form of origin stories, historical myths, or shared dreams of a desirable future, they are called 'social epistemes' in political science, collective representations of collectively perceived reality (Ruggie 1998, 184–6). Civilizations throughout history have been shaped by these socially constructed ideational overlays but because 'reality' also includes the material world, such representations typically co-exist with or in some fashion are grounded in their natural surroundings. For example, the language, stories, even religions, which inform a nation's character, will necessarily be coloured by technologies of the day (stone, copper, bronze, iron, steel, silicon), and by a people's economic relationship with the land (nomadic, subsistence, urban industrial). Robert Cox joins these factors in his definition of civilization as a "fit or correspondence between material conditions of existence and intersubjective meanings" (Cox 2002, 161).

Today, the correspondence between material and ideational aspects of the human experience is complete, but destructively so. The modern emphasis on material consumption and well-being, shared around the world, has been legitimized and accelerated by an overarching metanarrative called Progress and Prosperity. 'Prosperity for all' is the desirable end we seek, and progress driven by growth is the means to achieve that end. It is an alluring and generous story which has taken us not only to the very pinnacle of success, but also to the edge of calamity, foreshadowing a turbulent

DOI: 10.4324/9781032647067-5

climate, pervasive socio-ecological disorder, even an existential shock to modern industrial society. Instead of encouraging qualitative improvement of the human condition within the bounds of the available physical environment, our transcendent capacity for mindful oversight has aligned itself with and amplified our natural but naïve propensity to pursue material prosperity above all else, a process which inevitably enlarges the human impact on Earth and destabilizes the relationship between people and planet. Changing that metanarrative – or, more accurately, subsuming it within a better one – is the objective at hand.

I will show in the ensuing discussion that our 'transcendent capacity for mindful oversight' has been seriously compromised by its embrace of an aberrant conception of freedom, which contributes significantly to the amoral cast of today's hegemonic metanarrative. Freedom is and always has been a difficult concept in philosophy, in politics, and in economic matters too, but for present purposes we can rely on a straightforward understanding of it expressed in terms of 'freedom from' and 'freedom to.' In both cases, our moral compass has been lost. Freedom from oppression (variously defined) is clearly appropriate, but freedom from moral constraint, whether internally or externally imposed, is not. Likewise, freedom to choose and pursue a desirable future, and to fully explore one's potential, is also appropriate, but freedom to indulge the vision of a cornucopian future in which all wants will be indiscriminately satisfied is not. These latter, aberrant versions of freedom now colour and misdirect our shared mindscape, in large part because of our self-satisfied sense of exceptionalism and the resultant dislocation of our oversight capacity from a solid moral footing.

What is the origin of discursive epistemes? Karl Marx has suggested that "The ideas of the ruling class are in every epoch the ruling ideas" (quoted in Tucker 1972, 73). On this view, big ideas and dominant stories about social structure and behaviour, especially those that have taken root globally, are the self-serving creations of powerful actors who control the international political agenda through money, material resources and media ownership. This accords with the observation that social practices and categories of meaning are not value-free; that the process of discursive representation always privileges one type of knowledge over another. For example, the modernist (financial/industrial) era offers up a shared view of the world at the core of which is a secular, rational-scientific individual confronted by a single, empirically accessible reality. That view, now both common and exclusive, is contrasted to pre-scientific, pre-Enlightenment traditionalism, grounded in myth and superstition, while promoting instead the search for instrumental control of the real world and for material progress within it. Promulgated by those with the means and the interest to do so, a popular story such as 'modernism' can harden into the bedrock of common sense, the unquestioned merit of which prefigures society's moral orientation.

The linkage between formative ideas and powerful social actors is undoubtedly real, but the ideational domain is accessible by others too, and by other more subtle influences as well. The ideational world is ontologically distinct from the material world; a different set of causal influences is at play. One does not need a superior social position or the hammer and tongs of forcible coercion to construct a new discourse. Tools such as literature, art, emotion, critique, and philosophy are different from those used to physically dominate and direct the development trajectory of human society, and are available to anyone. A compelling discourse can originate anywhere and, because it is animated by symbolic (non-material) representations, it carries the potential to flip our shared mindscape from one mode to another virtually overnight. New stories about life, living, and fertile socio-ecological relations can have unique generative powers not beholden to venal influences, and as such they have the capacity to drive transformational change. For the moment, however, this potential remains unfulfilled.

A Short History

Neoliberal capitalism has been the *lingua franca* of policymakers around the world since the 1980s. This contemporary laissez-faire economic ideology is a modern manifestation of today's dominant metanarrative, but it is not the metanarrative itself. The real target for a radical makeover has deeper historical roots, namely, the broader, more encompassing story called Progress and Prosperity, an alluring and generous tale which promises more comfort, more security, more freedom and opportunity for everyone. Rife with half-truths, that story is on track to a bad ending. It has vaulted us to the very pinnacle of species success, but it has also seriously impoverished our living planet. Here is a truncated summary[1] of its (first European, then global) provenance.

It began at the feudal/modern juncture, some three hundred years ago, with the Age of Enlightenment and the ascendance of instrumental Reason.[2] "The Enlightenment was a magnificent revolution," said Antonio Gramsci. "It gave all Europe a bourgeois spiritual International in the form of a unified consciousness . . . In Italy, France and Germany, the same topics, the same institutions, and same principles were being discussed" (Gramsci, quoted in Hoare 1977, 12). Among other transformative events, it was this 'magnificent revolution' that separated the feudal and modern eras, and that gave birth to the seminal idea of progress.

Historian Ronald Wright defines progress as "the assumption that a pattern of change exists in the history of mankind (sic) . . . that it consists of irreversible changes in one direction only, and that this direction is toward improvement" (Wright 2004, 3). Of course this does not mean that human development before the Enlightenment was devoid of achievement. Pyramids and aqueducts, castles, trebuchets, and Gothic cathedrals were all significant accomplishments of architecture and engineering, and many other inventions such as the compass, gunpowder, and the printing press were also important contributions to the improvement of commerce, culture, and social development. But, as Ernest Tuveson puts it:

> None [of these various products] was thought of as 'progress' — simply as ingenious contrivances of persons, mostly anonymous, to meet immediate needs ... Lacking was the *idea* of progress; namely, the idea that purposive, concentrated effort . . . guided by increasing knowledge of nature, could achieve defined goals of improving man's (sic) estate in objectively measurable ways — and that such effort is of great moral, and indeed spiritual value. This idea, along with [democracy] has been the most enduringly influential force in the modern world.
> (Tuveson 1993, 515, original italics)

The transformation that pushed the idea of material progress to the forefront of public thinking was driven by a confluence of uniquely powerful historical factors. The Italian Renaissance of the

15th and 16th centuries, for example, foreshadowed the possibility of continuous progress by stimulating enormous advancements in painting and sculpture, literature and poetry, architecture, and music – in short, in wide swaths of cultural expression and knowledge. This was soon followed by the European Enlightenment which emphasized and encouraged the pursuit of rational and scientific enquiry, a necessary intellectual prerequisite for an era of rapid material improvement. The disintegration of feudalism, the rise of capitalism, and the Protestant ethic of personal responsibility, hard work, and thrift all spurred the process of capital formation and accumulation, even as new wealth poured into Europe from colonial empires. And, not least, the emergence of the international state system in the wake of the Thirty Years' War, with its emphasis on sovereignty, national wealth, and power, provided a conducive political context in which the foregoing were all given free rein. Sovereignty in particular institutionalized the notion of freedom, specifically freedom from external constraint or influence, and freedom to pursue any desirable future deemed in the national interest. These newly minted features of international political culture provided bedrock conditions for economic growth and paved the way for an emerging competitive market system which enthusiastically pushed progress forward.

The foregoing all merged to spawn the Industrial Revolution which boosted material productivity massively with machines and new sources of energy. Fossil fuels, especially coal, began to drive the machines of production in this new forward-looking society. The Revolution took off and consumer goods proliferated. Before, little changed from generation to generation. Now, cumulative changes became visible in ever shorter periods of time and improvements in living standards were enjoyed within the lifetime of an individual. The idea of material progress, finally brought within the perceptual horizons of ordinary people, could now be regarded as realizable and desirable. By the latter part of the 18th century, Adam Smith had expressed the widening consensus and the exciting new hope that "the progress of England towards opulence and improvement [would be] universal, continual, and uninterrupted" (Smith 1776, quoted in Arndt 1987, 11).

A radical shift in social perceptions (epistemes) ensued. Images of cyclical, recurring change gave way to the shared expectation of linear, straight-ahead movement to the future. This shift gave full voice to the Enlightenment ideal of self-determination, the personal freedom to pursue a better life. Within two centuries a collectively shared expectation of continuous progress had spread almost everywhere in the world, made manifest in the common pursuit of material prosperity in an ever-growing and more productive international economy. By the 1980s that expectation had been formalized in the neoliberal dream of a 'free market' unhindered by the drag of government intervention in the economy, a dream driven by the expectation of a cornucopian future in which all needs will be met, all wants satisfied. With restraints removed and horizons unbounded, progress is inevitable and prosperity is assured. Such is the promise of today's metanarrative.

It is easy to understand why prosperity for all has been selected as a desirable goal, from both a practical and a normative point of view. Practically speaking, we want everyone to be as well off as possible in order to reduce social tensions and to lubricate the steady flow of consumer goods through the world economy. It is a simple, appealing, and generous objective with obvious, tangible benefits. And from a normative point of view prosperity for all is fair as well as practical. It implies that the needs of the poor and disadvantaged will have been met and that a solid foundation of social equity will maintain and nurture healthy, productive relationships between all members of the world family far into the future.

It seems equally clear that progress has been chosen as the best way to eventually achieve prosperity for all. Regardless of how progress might be defined, certainly it has been closely associated with economic growth. A growing world economy can deliver the corporate surpluses necessary to foster research, innovation, job creation, and a continuous supply of consumer goods. And growth can also deliver tax revenues to support the development of the myriad public services a prosperous society needs and expects. In short, progress through growth can steadily improve living standards for everyone, and for next generations too.

Today, growth is openly and routinely highlighted as the paramount prerequisite for and driver of progress and prosperity; in fact, it is touted as the prime metric for almost all economic activity, whether at the level of the personal paycheque, the productive output of a corporation or country, and even global gross domestic product. Growth is measurable, universally coveted and deeply embedded in everyday discourse. For these reasons I take a closer look at economic growth as a key ingredient in today's dominant metanarrative.

Economic Growth

Growth is a process of two streams which, in keeping with the spirit of this book, may be characterized as material and non-material. The first stream is strictly quantitative. It refers to the production of more cars, more houses, more fridges, more food, and more factories. The second is qualitative – more comfort, more security, more freedom, more opportunity – a process of betterment which is often called 'development.'

Conceptually, the difference between quantitative (material) enlargement and qualitative (non-material) improvement is straightforward, but in practice it is not always easy to distinguish between the two, that is, between bigger and better. For example, more physical artefacts such as schools, hospitals, airplanes, and consumer goods all contribute to better living standards, and to the legitimacy and desirability of economic growth. There is in this respect an intimate and important connection between the real things we manufacture on one hand and our personal well-being on the other. Nonetheless, an unresolved tension lurks at the bottom of the relationship between the material and non-material, a tension which requires that the two be considered separately, each on its own merits. Specifically, *bigger* is a physical process limited by available space and resources, but *better*, a process of qualitative improvement, knows no such limitation. The growing totality of human artefacts on Earth will inevitably bump up against biophysical boundaries, but no such constraint limits knowledge creation, artistic and cultural expression, scientific advancement, or any other open-ended exploration of human potentialities. These two streams of growth, so closely identified one with each other, must eventually be distinguished from one another, and their relationship closely scrutinized.

The distinction is important because the subjective (non-material) and objective (material) modes of growth do not advance at the same pace. The latter moves forward easily. If resources are available, they can readily be transformed into (or purchased as) useful products. This process satisfies humanity's immediate need for things, not to mention our impetuous tendency to acquisitiveness, even gluttony. On the other hand, our self-reflexive ability to evaluate the quantity and form of consumption, and more generally to assess the utility of that consumption and its usefulness with respect to optimizing personal and social welfare, is much slower to develop, being more difficult, less tangible, and less gratifying in the short term. This is problematic because the oversight and evaluation capacity which rationalizes the process of responsible consumption originates from the subjective stream but, lagging behind, that oversight is weak and immature. In fact, today the subjective stream has been swept up in the rush to material prosperity; it has been co-opted to justify excessive behaviour. This suggests that the paired processes of growth and development need to be realigned and recomposed as a mutually beneficial duality which supports a balanced approach to social welfare, and which works to the protection of long-run planetary sustainability.

Another important and often misunderstood feature of growth is that it is not just the product of a capitalist economic order. Capitalism may be the most efficient and prolific producer of material things, and therefore a popular mode of allocation and production, but at bottom it is a means to an end (namely, growth-driven prosperity) and in that respect is not unlike other such modalities. Most other forms of socio-economic order (e.g. China's socialist market economy) are equally concerned with maximizing growth. Attacking capitalism as the chief cause of the global environmental crisis misses the deeper problem, that is, the physical impact of growth as such and, of equal importance, the worldwide political commitment to it, capitalist or otherwise, as a non-negotiable policy priority. The untouchability of the growth imperative demonstrates its deep embeddedness in the metanarrative of Progress and Prosperity, and the inviolate hegemony of that powerful modern episteme.

It is worth reflecting on this for a moment because the 'progress and prosperity' mantra is sometimes branded a Western phenomenon, and analysts who portray it as global in scope are regarded as failing to consider the life stories and hopeful expectations of any number of other local, regional, and national non-materialistic cultural narratives around the world. But this point of view, which sees the world through the 'diversity' lens described earlier, does not capture the larger picture which envelops all such non-Western social projects. This is so because governments, by and large, are excluded from the inventory of cultural actors, and yet it is government – the final arbiter of competing visions and the public face of any polity – that expresses the dominant bearing of any people (whether voluntarily or not) and which portrays that bearing to the rest of the world. When this crucial actor is included in the full comprehension and calculus of today's dominant metanarrative, the ubiquity of the commitment to growth becomes apparent; in fact, it is seen to be remarkably consistent, not to mention unique in history.

Without important exception, the growth paradigm encompasses all types of national government – capitalist, socialist, authoritarian, or otherwise – in the world's North and South sectors and remains constant despite variations in domestic policies which reflect divergent opinions about how best to realize it. Major institutions such as the United Nations, World Bank, International Monetary Fund, and World Trade Organization strive to optimize economic growth on behalf of member states. The G7 and G20, the OECD, and informal organizations such as the Trilateral Commission and the World Economic Forum, endorse and encourage growth. All regional trading blocs in the Americas, Europe and Asia were constructed to enhance the potential for growth. Multinational corporations pursue growth wherever and whenever possible as a charter responsibility to shareholders, and civil societies around the world are eager to fulfil their crucial role as consumers in a growing global economy. Together, all these institutional and social forces reveal, enhance, and drive forward the agenda of Progress and Prosperity.

In light of the curious fact that the pervasive commitment to economic growth permits and encourages the open-ended accumulation of material goods on an obviously finite planet, it is fair to ask how the idea of growth – which is to say, the belief that growth knows no bounds – came to be separated from the real constraints of the physical world. The evolution of that process is complex, but the truncated summary below will serve to provide some useful historical context.

Abstraction of the Growth Paradigm from the Material World

The disenchantment of nature – the rejection of myth, magic, and holism in favour of analytical reason and control – was the seminal event in the process of abstraction of ideas from the material world. The demotion of nature to a useful but often adverse thing, the philosophical separation of subject from object celebrated at the dawning of the Enlightenment, and the new pre-eminence of individual over society accented by the Protestant Reformation, all represented revolutionary epistemic change in Europe which granted priority to an anthropocentric worldview. These changes forcibly diminished the role of the natural environment even as they elevated the self-importance of people. But another more mundane change – the monetization of economic life – helped to usher in a market-based society which soon encouraged the commodification of all things tradeable, a process which contributed significantly to the evolution of dominant beliefs of the time, and to the disengagement of those beliefs from their material grounding.

The rapid development of a merchant class in towns across Europe spurred the process of monetization of economic transactions, spreading outward from urban developments, gradually encompassing more and more of rural society. Using money to facilitate trade and bargaining allowed the separation of use value (value derived from the actual use of real goods and services) from exchange value (measured by money) and, significantly, it opened the door to the theoretical possibility of unlimited accumulation of wealth. Food rots, water runs dry, buildings decay and people grow old, but money can be summed indefinitely and in perpetuity because it is symbolic, immaterial, and abstract.

The process of monetization soon extended far beyond the facilitation of trade of consumer goods and services to include the deep structural foundations of society and economy. Specifically, land, labour and capital were redefined as instrumental factors of production which could be treated as tradeable commodities, and with this important shift money and markets finally came to dominate the gestalt of social relations. Karl Polanyi clearly saw that, with this change, the economy, now a force unto itself, had come to depend upon a contrivance – an abstraction – because land, labour, and capital, their commodification notwithstanding, "are obviously *not* commodities; the postulate that anything that is bought and sold must have been produced for sale is emphatically untrue in regard to them" (Polanyi [1944], 1957, 72, original italics). Land, he argued, is not just property; it is the native soil on which an agrarian society is built. Likewise, labour is not just the paid work of employees; it is the daily life story of ordinary people. And capital, which includes productive plant such as warehouses and factories, may be understood even more broadly as that set of durable artefacts which collectively forms the heritage bequeathed to future generations. This definition includes both the built and the natural environments. Like productive plant, however, the natural heritage is also vulnerable to unpredictable investment flows, and may not survive the vicissitudes of business. Polanyi objected vigorously to all this. "Undoubtedly," he warns, "labour, land and money *are* essential to a market economy, but no society could stand the effects of such a system of crude fictions . . . unless its human and natural substance as well as its business organization was protected against the ravages of this satanic mill" (Ibid., 73, original italics).

Near the beginning of the 19th century the discipline of economics itself underwent a radical transformation from the classical to a neoclassical phase which added impetus to the process of abstraction of economic activity from social community and socio-natural interdependencies. That process began with the monetization of market relations and the commodification of factors of production, but it took firm hold with the pivotal decision to make the discipline of economics deductively rigorous and timelessly mathematical, an explicit attempt to emulate astonishing successes contemporaneously achieved in the field of physics. The classical understanding of economics as contingent and sociohistorical was displaced by a penchant for abstruse equations which forced economic management away from policymakers to an elite stratum of specialists whose technical analyses were now beyond the comprehension of non-experts. The forces of nature simply disappeared from these mathematical calculations, allowing the accumulation of capital to proceed irrespective of what was physically happening to soils, mines, forests, and waterways. Now floating free from material restraint, neoclassical economics did not impose any numerical upper limit on national economic growth, or by extension on gross world product, and present-day annual growth projections offered by the World Bank and the Organisation for Economic Co-operation and Development, among other organizations concerned with macroeconomic management and long-term trends, rely completely upon this theoretical licence.[3]

The next chapter will sharpen the critique of the incumbent metanarrative by interrogating the commitment to economic growth more closely, and by explaining its persistence. I will also sketch the main practical features indispensable to any new metanarrative concerned with the achievement of planetary sustainability. These features will be familiar to anyone with a lively interest in global environmental stability but are nonetheless worth rehearsing.

Notes

1 This summary is drawn from Purdey (2010, Chapter 3).
2 Some analysts trace the history of progress all the way to the ancient Greeks. For more on this, see Nisbet (1980).
3 The reader may see here an opportunity to apply Jürgen Habermas's Theory of Communicative Action (TCA) to the evolution of a new metanarrative. In TCA the Lifeworld is colonized by the System, blocking communication among participants. For Habermas, Lifeworld represents shared cultural patterns of meaning that enable actors to co-operate on the basis of mutual understanding, whereas System (instantiated typically as markets and bureaucracies) colonizes interpersonal cultural structures and processes by way of non-linguistic media such as money, institutional power, and formalized action programmes. The comparison between the socio-historical and neo-classical phases of economic thinking clearly invokes this Habermasian dichotomy and opens a rich avenue for the exploration and application of Critical Theory, especially if framed with regard to the important research agenda called 'decolonizattion.' Indeed, the corporate sponsorship of university buildings, financial awards and incentives, the selling of political parties like products on a shelf, and the portrayal of ordinary people as taxpayers instead of citizens all speak volumes about the depth of penetration of the System into the modern-day Lifeworld.

References

Arndt, H. W. 1987. *Economic Development: The History of an Idea.* Chicago, IL: The University of Chicago Press.
Cox, Robert W. 2002. *The Political Economy of a Plural World: Critical Reflections on Power, Morals and Civilization.* London: Routledge.
Gramsci, Antonio. 1977. *Selections from Political Writings, 1910–1920.* Translated and edited by Q. Hoare. New York: International Publishers.

Nisbet, Robert. 1980. *History of the Idea of Progress*. New York: Basic Books.

Polanyi, Karl. [1944] 1957. *The Great Transformation*. Boston, MA: Beacon Press.

Purdey, Stephen J. 2010. *Economic Growth, the Environment and International Relations: The Growth Paradigm*. New York: Routledge.

Ruggie, John G. 1998. "Territoriality at Millennium's End." In *Constructing the World Polity: Essays on International Institutionalization*, edited by John Gerrard Ruggie, 172–97. London: Routledge.

Smith, Adam. [1776] 1961. *The Wealth of Nations*, volume 1. London: University Paperbacks.

Tucker, Robert C. 1972. *The Marx-Engels Reader*, 2nd edition, edited by Robert Tucker. New York: W.W. Norton & Company.

Tuveson, Ernest. 1993. "Progress." In *The Blackwell Dictionary of Twentieth-Century Social Thought*, edited by William Outhwaite and Tom Bottomore. Oxford: Blackwell Publishers, Ltd.

Wright, Ronald. 2004. *A Short History of Progress*. Toronto: House of Anansi Press.

5 The Folly of Growth and the New Metanarrative

The stability, authority, and seeming inevitability – in other words, the sense of normalcy – conjured by any dominant paradigm can be shaken by the accumulation of anomalies generated by it. Several are now in play, not least of which is the massive problem of ecological overshoot (Wackernagel and Rees 1996; Rees 2020). The global economy has blundered across several 'planetary boundaries'[1] and, consequently, we are now draining Earth's natural capital (including biological diversity) to pay for the upkeep of present resource consumption. The inability to pay our ecological bills is a direct result of the fact that our self-reflexive ability to evaluate the quantity and form of consumption lags far behind our impetuous tendency to acquisitiveness.

Overshoot and an inability to evaluate quantity are both problematic anomalies of the growth paradigm related to its performance metric. Gross domestic product (GDP) is a direct offspring of the process of abstraction outlined in the previous chapter. This macroeconomic indicator, first deployed in 1934 after the Great Depression, embodies two seriously misleading properties, neither of which became apparent until the advent of the global environmental crisis. The first is that it conflates costs and benefits into a single category of economic activity. The cost of building a coastal seawall against rising waters, for instance, or cleaning up an oil spill, is included as a positive contribution to the (national or global) economy. This is exceptionally dangerous because it does not allow decision-makers to know when costs might exceed benefits, in which case further activity may be uneconomic. In contrast, microeconomic calculations at the level of the firm are

DOI: 10.4324/9781032647067-6

based on a clear division of costs and benefits. Specifically, when increasing marginal costs equal decreasing marginal benefits, business people know that further production of the firm's output would generate a financial loss. But no such rule applies to GDP. It is unique among economic indicators in its capacity to grow forever, growth which can occur simply because 'gross economic activity' includes any kind of transaction which can be measured in the abstract and limitless currency called money. The conclusion that GDP is in fact a poor measure of social welfare has been vigorously analysed for many years, yet GDP growth remains by far the most prominent and publicly recognized gauge of economic well-being. This speaks clearly to the sanctity of the worldwide sociopolitical commitment to growth, and no less to the illegitimate foundation of the metanarrative which it underwrites.

GDP growth also conceals a mathematical demon within it. In the public eye a rate of three per cent annual growth seems reasonable, just fast enough to accommodate a rising living standard for a growing population. It is a rate that, if sustained over a period of time, would please most political leaders. And yet, seemingly unknown and certainly unremarked, this slow and steady growth rate of three per cent per year is in fact an exponential function – and if we have learnt anything from Thomas Malthus, we have learnt that exponential functions are explosive. The overlooked mathematical reality here is that fixed annual increases are being applied to an ever-larger economic base, and this simple fact leads to a remarkable result. Leaving the mathematical details aside, one finds that an ever-larger base, steadily pushed, will double in size at regular intervals, as in the sequence 2, 4, 8, 16, 32, 64, and so on. In the case of a constant three per cent per year increase, each doubling occurs approximately every 25 years (the doubling interval for ten per cent annual growth is seven years). Even allowing for various dampening effects (such as periodic recessions and setbacks like viral pandemics), this trajectory points to an increase in the size of the global economy from US$96.5 trillion today to perhaps US$150 trillion by 2050, and then US$250 trillion by 2075. In light of the fact that we are already in a chronic condition of ecological overshoot, our intention to generate such a stupendous increase in economic activity is imprudent, to say the least. And yet, political and corporate leaders prefer instead to believe in the boundless capacity of human ingenuity to overcome any obstacle.

To secure this point, they often point to the hapless Malthus as the archetypical pessimist.

In the eighteenth century Thomas Malthus compared the rate of population increase in England (an exponential function) with the rate of agricultural production (a much slower linear function of the form 1, 2, 3, 4, 5 ...) and concluded that population growth would soon outstrip food supply and countless thousands would perish of starvation. This did not happen for a variety of reasons, chief among them huge increases in agricultural productivity. Malthus was subsequently savaged for his fear of an obscure mathematical menace and his embarrassing underestimation of humanity's ability to successfully meet any kind of challenge. The modern-day epithet 'neo-Malthusian' is still reserved for those whose pessimistic outlook is derided as unsubstantiated alarmism.

Blindly optimistic but at least marginally responsive to popular demands for a safe and stable environment, how exactly are today's managers planning to minimize the impact of a growing world economy on vital resources and ecosystems? The remedy proffered is 'green growth' whereby steady reductions in resource intensity and pollution per unit of economic output will eventually overtake increases in scale of production. Reduced resource intensity and declines in pollution (especially carbon dioxide) are obviously good things, but such gains are of little use if ever-increasing production continues to overwhelm efficiencies per unit of output, as it does today. Decoupling growth from its physical impact in absolute terms is very hard to accomplish. To be truly effective any transition to 'green growth' would have to occur rapidly enough to render present activity environmentally benign, to pay down our burgeoning ecological deficit, and to accommodate the surge in future activity commanded by the exponential function. The expectation that economic activity can be decoupled from the environment quickly and completely enough to accomplish these critical objectives beggars belief, and yet that expectation lies at the heart of today's mainstream argument in favour of continuous economic growth.

The Persistence of Growth

Serious anomalies associated with the process of economic growth as the driver of progress and prosperity have not by any means been persuasive enough to de-legitimize it as the prime policy directive of political and corporate leaders everywhere. This is anomalous in itself and requires some explanation. Three factors explain the persistence of the commitment to growth: political expedience, moral convenience, and a distorted conception of the Enlightenment ideal of freedom.

The story of Progress and Prosperity serves up an image of the future that is equitable, cornucopian, and ecologically benign. North–South issues will be resolved by integrating developing and transitional economies into the growing global market, allowing impoverished states a progressively larger slice of a progressively larger pie. Future generations will benefit from the process of decoupling the global economy from its environmental impact, and their heritage will be secured with the production of capital goods which replace any resources exhausted in the present. And non-human species and their habitat will be protected by the moral beneficence engendered by prosperity.

In addition to delivering this idealized image of the future it has also been said that growth – the operational foundation of today's metanarrative – serves present political interests because it is the only policy priority which penalizes no one. All contending parties, public or private, employer or employee, rich or poor, can agree on economic growth as a policy objective simply because it benefits them all by promising more of everything for everybody. For this reason, above all others, John Kenneth Galbraith has concluded that

> No other social goal is more strongly avowed than economic growth. No other test of social success has such nearly unanimous acceptance as the annual increase in the Gross National Product. And this is true of all countries developed or undeveloped; communist, socialist or capitalist.
>
> (Galbraith 1968, 183)

Contemporary rulers who preside over prosperous economies are likely to enjoy power, social stability, and electoral success, and those who are ruled are likely to enjoy enhanced prospects for personal security and steadily increasing opportunities for improved standards of living. If prosperity lags, the growth paradigm promises a better tomorrow, a promise on the strength of which current problems, shortages, or challenges may be more easily borne. In this way the paradigm is perceived not only as desirable and useful but also, in light of its progressive teleology, as the right way to improve the human condition. Governments' common pursuit of growth strengthens their hold on power even as it serves the will of the people, which in turn affirms the legitimacy and rectitude of their policymaking.

In concert with the advantages of political expedience, growth is ostensibly the best of all social lubricants too, which is to say, it is morally convenient because it relieves the pressure of tensions associated with inequality and competing claims on limited resources. In *The Affluent Society* Galbraith observed how "increased production [had become] an alternative to redistribution," that it was "the great solvent of the tensions associated with inequality," and that "it is the increase in output . . . not the redistribution of income, which has brought the great material increase, the well-being of the average man (sic)" (Galbraith 1958, 86–7). Growth has displaced the moral injunction to redistribute resources between rich and poor, to husband them for the benefit of future generations, and to share them between humans and other life forms on Earth. By facilitating this moral defection without incurring the revolutionary wrath of civil society, the paradigm binds rulers and ruled together in common cause.

Political expedience and moral convenience are powerful attributes of the paradigm but what makes growth the right thing to do – what makes growth normatively legitimate – in the eyes of the public is its close association with political and economic freedom. Progress, as we have seen, is made manifest in the material benefits generated by economic growth but it has also, from its inception at the Enlightenment, included a notion of freedom – freedom from superstition, domination, oppression and exploitation, and freedom to independently pursue desirable goals and a better life. Thus, progress has always comprised both a quantitative and a qualitative dimension; the two are mutually involved and, arguably, progress is best achieved when these dimensions are balanced or harmonious. An increase in knowledge, for example, will be most salutary if it engenders or facilitates improvements to the physical security and material welfare of people, or to the ecological viability of Earth. A process of abstraction, however, traced earlier, has de-linked the qualitative (ideational) aspect of progress from the quantitative (material) aspect of growth such that the mutual relationship between them has been radically disengaged. The belief in progress and in the expectation of continuous improvement of the human condition is no longer accountable to material limitations. And the notion of freedom, embedded in the idea of progress but now also relieved of accountability, has been reimagined to mean freedom to pursue a limitless and cornucopian image of the future.

The hard reality, however, is that unrestrained freedom is a distortion of Enlightenment ideals. Such freedom is illusory if it ignores the limitations of Earth's finitude. It is reckless if it encourages a cornucopian vision of the future in which human betterment is defined in terms of continuous consumption. And it is licentious if it releases decision-makers from the need to make ethically informed choices which may constrain growth-driven behaviour in favour of the larger, longer-term common good. False freedom and the moral laxity it engenders are the poison pills of the modern story of Progress and Prosperity.

The Persistence of Growth, Continued

The reasons just proffered for the persistence of growth are, in a manner of speaking, theoretical, but these ideational attributes are congruent with two concrete facts which add both substance and danger to the issue, facts best expressed by the image of humanity having dug itself into a very deep hole.

Our shared belief in a bountiful tomorrow, and in the generative power of economic growth, has encouraged us to literally mortgage the future in order to maximize well-being today. World GDP now stands at approximately US$96.5 trillion, but world debt, which supports the lifestyle we enjoy now, dwarfs this number at US$303 trillion. The expectation is that this enormous debt can and will be paid down in due course by growing the economy hard enough and fast enough to generate the surplus necessary for this purpose. If we rule out alternative repayment solutions such as globally imposed austerity measures or some novel kind of re-set or write-down, then we have committed ourselves to an aggressive pattern of growth until the books are balanced.

The second substantive issue has been created by the world's oil industry. The current consensus regarding climate change appears to be that emissions of carbon dioxide must be quickly and radically curtailed, that a significant percentage of greenhouse gases already in the atmosphere must be scrubbed out, and that any new energy produced by burning fossil fuels must be carbon-neutral. Given the abysmal failure of carbon capture and storage technologies to date, this latter requirement cannot realistically be met in the foreseeable future, prompting the familiar cry of 'Leave it in the ground!' From an oil company's or petro-state's point of view, however, this suggestion is laughable; too much potential profit is at stake. But, far worse is the fact that recoverable oil reserves around the world have already been monetized, meaning that oil-in-the-ground has been treated as a live asset appearing in stock valuations and shareholder portfolios, in security for loans, and in other locked-in corporate holdings. Leaving oil in the ground would generate 'stranded assets,' the total value of which is currently estimated at some US$20 trillion.[2] This is another unfortunate example of mortgaging the future premised on the expectation of continuous growth. The monetization of oil reserves and world debt in general both provide tangible and plausible explanations for the persistence of growth. A generous account of these problems would suggest that they result from the kind of naïve exuberance and carefree optimism typical of an adolescent mindset. A less generous point of view would suggest that a serious social pathology is at work, that adulthood is nowhere in sight; that both the ideational and material forces which have endangered our future by misdirecting our way forward must be confronted directly.

The New Metanarrative: Some Practical Considerations

The new global narrative is still incubating and it is not clear yet how its distinctive, radically transformative features will emerge or be expressed. In the meantime, there are a number of ways to imagine the contours of the narrative which give it a general shape. Four such examples follow. First, the narrative may afford an opportunity to examine manageable sub-components of transformative change independently.

Disaggregating Transformative Change

Thomas Homer-Dixon has offered a useful parsing of change into four sub-categories (Homer-Dixon 2009), each of which directs our attention to specific themes which might be embedded in the new story:

a. *Cognitive transition*: This entails a shift in dominant worldviews from Newtonian mechanics to complex adaptive systems (CAS). Newtonian mechanics inform the Machine Age, but planetary operating systems, ecosystems, human society, and the global economy are all CAS. Different causal forces are at play and new phenomena, such as non-linearity, resilience, and, especially, emergence become salient.

b. *Economic transition*: This entails a shift of the global economy from the growth model to the steady-state model. In the former, the environment is a subset of the economy. In the latter, this is reversed: the economy is a subset of the environment.

c. *Political transition*: This entails a move towards a more muscular, informed, and engaged civil society which can successfully hold political leadership to account. Modern communication technologies can be harnessed for this purpose.

d. *Normative transition*: This entails less emphasis on utilitarian ethics and moral relativism, and more on existential issues and spiritual values, prompting deeper reflections on metaphysical questions.

These component parts of transformative change clearly indicate the multidimensional nature of the sustainability challenge, and the layered look of the new story.

Three Circles

In 1987 the United Nations published *Our Common Future*, also known as the Brundtland Report. This report brought the phrase 'sustainable development' into mainstream culture, presenting a vision of a secure world resting on three equally important tenets: environmental protection, economic growth, and social equity. It was tremendously valuable as a conversation-starter but now it is dated and in some important ways, wrong. A better approach to the new story starts from the premise that these tenets are not equal partners; that transformative change will fundamentally restructure the interactive relationships between economy, society, and the environment. A corrected image of sustainable development changes three equal tenets into three nested circles: economy *within* society *within* environment.

Sustainable development was, in its original formulation, a compromise. 'Sustainable' was directed at the Global North, intended to constrain excessive consumption and pollution, while 'development' was a stand-in for economic growth, a concession to and tacit permission for the Global South to industrialize quickly in order to redress North-South economic disparities. It was the prototypical example of an environment versus economy compromise but, despite economy and environment being ingenuously portrayed as equally important, it is now abundantly clear that when pressure mounts the real power of economic stakeholders is bared and environmental concerns are quickly sidelined, thwarting the admirable objective of protecting the needs of future generations.

Notwithstanding the evident power of economic forces, however, recall that economic relations are the invention of, are embedded within, and are (or should be) fully accountable to the community which constructed them for the specific purpose of providing affordable goods and services to people. The economy is a tool, a means to an end. Managerial oversight is required not only to ensure a steady stream of goods and services but also to guarantee that economic activity does not compromise the ecological life-support systems on which society depends absolutely for its sustenance and longevity. Economy and society are not equal partners.

Nor are society and the environment equal partners. Environmental protection is clearly a social responsibility, but it is not a responsibility defined in strictly anthropocentric or instrumental terms. The environment is not a means to an end; it is not just a source or a sink made available for our convenience. We now know that society – here meaning the human population in its entirety – is embedded within and wholly dependent upon the natural Earth. Humanity is an evolutionary product of the environment, not its maker or its master. Relations between social and environmental circles are complex and interactive, but they are not in fact equal; they are hierarchical. On this view, then, contextual imagery places society within the global environment, just as the world economy is placed within human society.

The three-circle imagery is more than a structural depiction of relationships which would obtain in the wake of transformative change. It also speaks to basic functions which each of these components would undertake. The inner circle (the world economy) will still be charged with delivering affordable goods and services to people everywhere by efficiently allocating resources and nurturing innovation even as it eschews physical enlargement (quantitative growth) beyond the capacity of Earth to support it. The middle circle (human society) will set operational parameters for the world economy which prevent economic activity from overrunning the biosphere, and sociopolitical oversight will mitigate extremes of wealth and poverty. The outer circle (the bio-geosphere) places limits on the maximum size of the human population on Earth, and on the total amount of energy and material that can be diverted for economic purposes. Population size and per capita resource consumption may vary according to changeable definitions of sufficiency or the good life, but in general limits imposed by a finite planet will constrain the number of people who can enjoy a reasonable standard of living during any given period of human history. The new global narrative will undoubtedly allude to if not directly incorporate imagery as described above, and it will reflect a new and more relevant definition of sustainable development.

Rules of the Road

In order to establish a clear sense of direction, the new global narrative will delineate the immutable survival conditions which must obtain under post-transition circumstances, expressed as rules which follow directly from the section above. These rules are already well known, at least to ecological economists. They form the bedrock conditions of a steady-state world economy. They are axiomatic and require no further elaboration here; in some fashion they will be indispensable to the new story.

a. The use/consumption of renewable resources will not exceed the ecosphere's regenerative capacity.
b. The use/consumption of non-renewable resources will not exceed the rate at which substitutes (where possible) can be developed.
c. Waste generated by human society will not exceed Earth's assimilative capacity.
d. The human population on Earth will be (more or less) stable.

e. The built environment will be durable.
f. Our population and built environment will be maintained by a minimal level of (low entropy) energy/material throughput, consistent with a good standard of living for everyone.

All these points are informed by the question of scale, which is to say: how big exactly can the human presence on Earth be without compromising our own well-being or damaging the planetary ecosphere? The exact parameters of optimal scale are in fact unknown at present, but Daly and Cobb make clear that a precise predetermination of them is not necessary. We need only bear in mind that a steady-state economy, however large, should be sustainable in perpetuity at levels of per capita resource use that permit a good life for all. Knowing this,

> Our first goal therefore should be to stabilize at existing or nearby levels as soon as we reasonably can. Once we have learned to be stable at some level, then we can worry about moving to the optimal level … If we do not know how to be stable, then identification of an optimal scale will only allow us to recognize and wave goodbye to it as we grow beyond it … Those who argue that there is no point in talking about stability unless you can first specify the optimal scale … have got it backwards. *Unless we are willing and able to be stable, there is no point in knowing the optimum.*
> (Daly and Cobb 1994, 241–2, italics added)

Complex Adaptive Systems

In the subsection above labelled 'Disaggregating Transformative Change' I mentioned that the move to planetary sustainability comprises several interrelated transformations, not least of which is a shift in worldviews from Newtonian mechanics to complex adaptive systems (CAS). Newtonian mechanics inform the Machine Age but planetary operating systems, natural ecosystems, the global economy, and human society at large are not machines. They are complex adaptive systems in which different causal forces are at play and new phenomena such as resilience, non-linearity, and emergence become salient. A successful transition to sustainability will require the competent management of these forces and phenomena, a task for which a new analytical framework will be indispensable.

The largest CAS on Earth is composed of two parts, namely, the atmosphere and the world ocean. The atmosphere weighs 5 quadrillion tons and occupies 1.2 trillion cubic kilometres. The world ocean adds another 1.5 billion cubic kilometres of volume, and 10 quintillion tons of weight. The two parts of this unitary system are dynamically coupled, exerting huge forces on each other. Trillions of tons of water are exchanged between them every day. Powered by heat from the Sun, this massive, moving system is the engine of Earth's climate; indeed, it *is* Earth's climate.[3]

The next largest CAS on Earth also has two major components. The first part is the planetary ecosphere and the second part is the population of human beings embedded within it. Together they make up what we now call the socio-ecological system. Like the atmosphere–ocean complex, the two parts of this CAS are deeply interactive, giving rise to a bewildering array of changes large and small, some instant, some glacial. The human population, growing at some 70 million people each year, is changing Earth's physical landscape massively and permanently, altering ecosystems and disturbing the lives and livelihoods of other species. In turn the planet changes our behaviour as we adapt to land, air, and waterways altered by human intervention. The combined entity of people and planet – that is, the planetary socio-ecological system – is enormous and dynamic, constantly adapting, moving, and evolving.

The major source of uncertainty in the relationship between people and planet in the past was the changeability of nature, but today, because of growing stresses imposed on Earth's operating systems by human activity, the dominating cause of unpredictability is the macro-behaviour of the human species. We are changing the composition of the atmosphere and ocean with little comprehension of the consequences. Fresh water systems, soils, forest cover, species habitat, and much more are all vulnerable to roughshod human intervention. Given this new source of unpredictability it follows that human activity – the disruptive element in this relationship – should be constrained. Managing the relationship between people and planet, and consequently the evolution of the planetary socio-ecological system, is tantamount to managing human behaviour. Newton's laws tell us nothing about how to accomplish this, but a better understanding of CAS may be helpful.

Complexity is a rich source of innovation, and it is also a source of stability. Diversity among its parts, distributed capabilities among its component sub-systems, and redundancy to protect against failure all combine to make CAS highly resilient to disturbance. Too much complexity, however, can cause vulnerability. If connectivity is too dense, if component parts are too tightly coupled, if a breakdown here spontaneously creates a breakdown there, then overall system resilience decreases. The ability to adapt becomes brittle instead of flexible, and surprising changes in behaviour can be shocking and damaging instead of useful.

Arguably, the tightly coupled connectivity so characteristic of modernity – in our financial markets, in our food production and delivery systems, in our transportation and communications networks, in our electrical power grids, and much more – has made society too complex. All these sub-systems serve us well, but they seem poised on a knife-edge, vulnerable to catastrophic failure. The management lesson from CAS theory is that the resilience of modern industrial society, that is, the capacity to survive disturbances, should be improved. This means decentralizing the dense nodes which control these sub-systems, distributing capabilities across a richer and more varied socio-industrial landscape. It means easing our locked-in dependence on far away suppliers of goods and services, and on irreplaceable technologies understood only by an elite few. Increasing resilience means, to a large extent, increasing self-sufficiency and autonomy.

In terms of social organization, the need for resilience calls us to envisage the democratic empowerment of local communities in order to increase diversity, redundancy, and novelty; to gain greater control over basic services vital to our lives; and to reduce our ecological footprint. This devolution to local empowerment is both necessary and important, but it misses a significant point. Local autonomy is direct and practical but, if the move to planetary sustainability is to succeed, there must be some degree of cooperation and coherence among these communities, and some overall sense of direction for their collective evolution as well. Recognizing the human population on Earth as a CAS interacting with its material environment leads to a new emphasis on decentralization, but by the same token it amplifies the need for supranational coordination. It is a central contention of this book that the purpose and directionality of that coordination is profoundly influenced by the collective ideational experience of human society at large.

Notes

1 For more on this, see Stockholm Resilience Center at www.stockhol mresilience.org/research/planetary-boundaries.html
2 For a summary of this argument, visit the Capital Institute at https:// capitalinstitute.org/blog/big-choice-0/. Note that US$20 trillion compares with the cost to the world economy of about USD12.5 trillion caused by Covid. See www.reuters.com/business/imf-sees-cost-covid-pandemic-rising-beyond-125-trillion-estimate-2022-01-20/
3 The sheer bulk, energy, and sensitivity to change of the atmosphere/ ocean complex make it exceptionally dangerous if carelessly prodded.

References

Daly, Herman E., and John B. Cobb. 1994. *For the Common Good: Redirecting the Economy Toward Community, the Environment, and a Sustainable Future*, 2nd edition. Boston, MA: Beacon Press.

Galbraith, John Kenneth. 1958. *The Affluent Society*. London: Hamish Hamilton.

Galbraith, John Kenneth. 1968. *The New Industrial State*. New York: The New American Library, Inc.

Homer-Dixon, Thomas. 2009. *The Great Transformation: Climate Change as Cultural Change*. Conference presentation, Essen, Germany, June 8. www.homerdixon.com/2009/06/08/the-great-transformation-climate-change-as-cultural-change/

Rees, William E. 2020. "Ecological Economics for Humanity's Plague Phase." *Ecological Economics* 169. https://doi.org/10.1016/j.ecolecon.2019.106519

Wackernagel, Mathis, and William Rees. 1996. *Our Ecological Footprint: Reducing Human Impact on the Earth*. Gabriola, British Columbia: New Society Press.

Part II
Morality and Agency

6   An Introduction to Dualism, Monism, and the Problem of Reconciliation

I walked out of Africa and peopled the world.
I built civilizations that lasted a thousand years.
I built monuments to the gods I worshiped.
I built monuments to myself.

There is no question that we humans are an extraordinary species. Whether we are an exceptional species, that is, ontologically unique compared to all others, is a closely debated question with no compelling answer in sight. Are we 'ensouled' where other animals are not? Do we enjoy a kind of morally significant capacity that grants us superiority over non-human beings? Or, more prosaically, do our worldly accomplishments such as space travel or instantaneous communication or symphonic productions make us special by definition? The debate is likely to continue but in the meantime certain facts intrude which may render this an unimportant semantic tussle. After all, our status on Earth speaks for itself: we as a species are an apex predator or, more accurately, *the* apex predator. We rule the planet – ruthlessly, some may say – and this reality makes the distinction between exceptional and merely superior, or dominant, moot.

Indeed, a persuasive argument can be made that we are not exceptional at all, that in terms of population dynamics our collective behaviour is indistinguishable from that of any other living entity which enjoys free access to all resources, and which no other animal can successfully resist. That kind of behaviour, however, that kind of overwhelming power, also comes at a cost. Success breeds failure; as resources are eagerly consumed and population growth is freely indulged, a crash is inevitable. The 'boom and

DOI: 10.4324/9781032647067-8

bust' dynamic in such situations is common and well-documented, whether it applies to deer without wolves on a leafy island, bacteria free to grow only to expire at the edges of a nutrient Petrie dish – or, more ominously, cancer cells that reproduce exponentially until the host organism is killed. These comparisons are of course odious and demeaning and, one hopes, unrealistic. And yet, eminent ecologist William Rees has concluded unambiguously that human numbers on Earth are now pushing through the 'plague phase' which, as in all other similar cycles, immediately precedes a population bust (Rees 2023). From this uncomfortably deterministic point of view we as a species may be extraordinary, but we are certainly not exceptional or morally superior in any important way. In fact, because our macro-behaviour is indistinguishable from those examples just mentioned, and many others like them, we are for all intents and purposes identical to them (according to the *Identity of Indiscernibles,* an ontological principle first formulated by Wilhelm Leibniz), meaning that in a real and very unfortunate sense we are not distinctively human at all. As Rees puts it: "[I]f the world's nations cannot come together to fully engage their common fate, humanity proclaims itself to have no more practical intelligence or conscious moral agency … than does any other species in overshoot at the brink of collapse" (Ibid., 19).

It is ironic, perhaps even paradoxical, that life will extinguish itself if its exuberance and fierce drive are allowed free rein in a world otherwise constrained by edges, boundaries, and thresholds, yet that is precisely what happens. It may be that the root of exceptionalism, or at least an important aspect of it, and that which ultimately distinguishes us from others, is the ability to voluntarily and pre-emptively impose limits on self-behaviour, for the sake of self-interest to be sure, but also in recognition of and respect for the contextual elements that delimit that behaviour. If this reading of exceptionalism has merit, it may be premature to abandon any such notion as 'moral agency,' founded as it must be on foresight and care; premature to ignore the possibility that it can be cultivated and enlivened, not beholden to blind biological impulses. From what source that agency might arise is difficult to discern, however. If it is true (as I will argue in due course) that agency is an implicit potentiality only in the lived experience of individuals, then individuals may be called upon to provide the critical counterweight to the burden of insensate collective behaviour.

The foregoing introduces themes of morality and agency. Part I of this book dealt primarily with the pragmatics of constructing a new metanarrative. Part II will focus more on the ideational underpinnings of that construction project in the interests of securing the ontological foundations of both agency and morality. Agency, the toughest nut and *sine qua non* of planetary sustainability, will be discussed mostly in Chapters 9 and 10. An exegesis of morality will run through Chapters 7 to 10. I will begin Chapter 6, however, by continuing the discussion of dualism, a theme now familiar to the reader and alluded to above in the tension between individual and collective roles in the macro-behaviour of human society on Earth. This theme is of central importance for two reasons. First, the successful evolution of a new metanarrative depends on the reconciliation of the material and ideational aspects of the sustainability problematic, a process that inevitably conjures the mind–matter duality. And second, the general approach to reconciliation undertaken here is an emulation of Aristotelian *phronesis*, the pursuit of practical-moral knowledge, which is to say, the pursuit of moral knowledge in the context of particular worldly circumstances, and how that knowledge might be ascertained. In terms of the environmental crisis, climate change is the most pressing of these worldly circumstances. And, because existential threats prompt, or should prompt, deep reflection on meaning of life issues, the pursuit of moral knowledge will be introduced below with an impressionistic foray into theology. I will pursue a speculative discussion of 'godhood' in subsequent chapters.

Dualism

The distinction between objective and subjective points of view, between explaining and understanding, between the quantitative and the qualitative, between positivism and *verstehen*, between *poiesis* and *phronesis* has divided the social sciences and humanities since antiquity. Generally speaking, the latter has tended to involve the interrogator in whatever exploratory exercise is being undertaken. This involvement, by bringing a personal element into the process, is therefore necessarily idiosyncratic even though, its proponents argue, no less rigorous, no less useful. On the other hand, striving to find truth by separating the agent from the process, by eliminating individual bias, is surely an admirable goal which clearly answers the call of reason; and in fact within the narrow confines of the Newtonian world examples abound of technical and mechanical processes, concisely repeatable in the lab and in the field, whereby objectivity appears to have been successfully realized. That these processes follow known, universal laws of physics supports the argument that, by extension, all of reality should follow those same laws. This happy expectation, however, is frustrated by the awkward truth that none of the social sciences nor any of the humanities have identified even a single inviolate law by which the future behaviour of individuals or groups of people might be reliably predicted. People, evidently, have a will of their own, unanswerable to science or reason.

It seems to be unavoidably necessary to posit two incommensurable domains, the material and the non-material, as Descartes certainly did, but this distinction has provoked a host of questions which arise if an unbridgeable duality is posited; for example, how exactly do these two domains interact, what kind of relationship do they have and how, if they are incommensurate, can one exert any influence on the other? No consensus on these issues has been reached, leaving some interlocuters to suggest that the debate is specious, that at bottom there must be only one kind of reality. With this hypothesis in mind, those with a propensity for materialism aver that all 'ideas' are just manifestations of physical phenomena, just neurons firing in the brain. Idealists, on the other hand, argue the contrary, namely, that because the 'real' world we engage with is only perceptible by way of our senses, it is therefore ultimately inaccessible with no objectively knowable physical foundation. Neither side has persuaded the other, so the materialist/idealist divide remains unresolved.

Materialism and idealism both stake claims to universality but, arguably, the inability of each to dissuade, defeat, or subsume the other suggests that neither claim meets the standard of all-inclusiveness. What's really called for here is a deeper kind of 'monism,' that is, a theory that posits an underlying unity which joins all into one, and from which our perception of a fragmented world emerges. And, in fact, it turns out that monism is a philosophical concept with an ancient and distinguished pedigree. There are multiple religious doctrines which posit an aboriginal entity which is expressive of, or, more accurately *is*, the living foundation of all of reality. The Hindu Brahman is a clear example of this. The Brahman is described in Indian sacred writings called the Upanishads as the supreme existence, absolute reality, the creative principle that underlies the diversity of the world. Religion thus understood merges pluralism into a single proto-phenomenal entity which, in the West, we would call God.

Chinese Taoist philosophy offers a (familiar pictorial) representation of monism in which opposite forces (yin and yang) are complementary pieces of a larger, black and white, whole.[1] Likewise, the Egyptian god Isis, the 'mother and father of all things,' embodied an all-encompassing unity, and in Greece Pythagorus, Parmenides, and Plato all saw the universe as a single entity while Heraclitus opined 'From all things One, and from One all things.' In modern times, Romantic poets, Mozart, and scientists Newton, Faraday, and Einstein all added their voices to a call for a monistic philosophy. Throughout, the Roman Catholic Church opposed this historical theme, arguing that God and nature were separate entities, and that any effort to conflate the two amounted to a demotion for God. This point of view has largely prevailed despite the possibility that this conflation may not entail a demotion of God, but actually a promotion of nature.

More recently, something called 'neutral monism,' a phrase originally coined by Bertrand Russell, posits a putative unity without going so far as to suggest anything truly universal. Neutral monism suggests a kind of originating condition which precedes both mind and matter, from which both emerge; it prioritizes neither one nor the other and in that sense is 'neutral.' This theory too, however, has run into difficulties because proponents of this point of view, including Russell, William James, Ernst Mach, and others, have proposed a plethora of types of neutral monism all of which co-exist simultaneously in the world, but none of which lays claim to all-inclusiveness, including such things as experience, sensations, precepts, and various kinds of generalizations such as 'information'; and, unsurprisingly, there remains an unclear understanding of how exactly these monisms are related to each other or to their dualistic progeny, and how they interact with them. This theory, in short, is incoherent.

Yet another kind of oneness has been hinted at by contemporary quantum physics, which argues for a 'superposition' from which the classical (perceptible) world emerges, and for sub-atomic 'entanglement' that joins disparate particles and processes together which otherwise would be unable to communicate with each other. The actual existence of these mysterious phenomena has been well documented and, despite being poorly understood, they promise an intriguing interpretation of the world. The practical usefulness[2] of these manifestations notwithstanding, however, the 'All is One' mantra is perceived in the main by science to be a bridge too far, disdained as quasi-religious and not useful analytically; it doesn't solve problems and is therefore left to the airy ruminations of philosophers.

I adopt in this book the holistic interpretation which asserts that the universe (all that exists) is of a single piece comprising within it a range of types of being. In the beginning, when an aboriginal unity was fractured into parts and began its journey through time and evolution (as discussed in subsequent chapters), duality (the simplest form of many) was the first thing to come into being and in that sense is a primordial feature of reality. Significant (and often baffling) dyads range from, at the very foundation of reality, energy/information, the wave/particle duality familiar to physicists, up the compositional ladder to the inanimate/animate distinction which ostensibly separates life from all else, to the subjective/objective division which purportedly isolates observer from observed, and all the way up to the deep conflict between conscious, self-reflexive awareness and the natural Earth, the tension which stands at the core of the global environmental crisis. For present purposes, this latter duality may be understood as two different types of entity which are ontologically unique but not mutually exclusive; they do not constitute or inhabit separate and incommensurable worlds.[3] On this view, mind and matter co-exist within a naturally whole continuum. They are, however, under no obligation to co-exist in harmony or to their mutual benefit and, as discussed previously, they do not. Their current concordance is in fact, with respect to sustainability, counterproductive. The physical foundations of life on which we depend completely have been negatively impacted by today's dominant metanarrative, a story that encourages materialism and legitimizes impulsive behaviour. That metanarrative is evidently incapable of restraining or otherwise guiding the unreflective behaviour prompted by easy access to the benefits and conveniences of the material world. Reconstituting the relationship between mind and matter is key to resolving the sustainability challenge.

If the universe is of a single piece, then the relationship between mind and matter cannot be as dichotomous or as simple as the former sensing the latter, the latter providing material grist for the former. There must be a deeper, perhaps more contemplative union between the two which expresses or embodies the underlying unity in which they both participate. There is no unambiguous way to ascertain conclusively what form this 'contemplative union' might take, so an interpretive point of view, and the impressions it generates, seems more apposite. Impressions, providing that they comport with reason, accessible empirical observations, and other appropriate validation criteria, are evidently an inescapable feature of qualitative inquiry.

Climate Change and God

'Impressions' are colloquial versions of the logical and reasonable conclusions we reach about the nature of reality; they provide a subjective overlay on the brute facts of the world. One does not have to embrace the extremity of idealism to appreciate that personal impressions form an integral part of the human psyche, and of our collective experience of the world. Given that we have no direct access to or categorical knowledge of the material world (as I will make clear in the next chapter) or, for that matter, to the perceptual tools we use to probe it, impressions are both unavoidable and indispensable. They are also, sometimes, wrong, or misleading.

Take, for example, the fact that nature is not really green. Chlorophyll is the molecule in plants that absorbs light from the Sun and this energy in turn powers the all-important process of photosynthesis which produces food for the plant. Chlorophyll absorbs light from both the red and blue ends of the visible spectrum, but not from the middle green wavelengths. This colour is not useable in photosynthesis and is therefore rejected by chlorophyll, bounced back out into the world where we gather it into our eyes and perceive it as, of course, green. That plants appear green even though, essentially, this colour is anathema to them, is deeply embedded in our factual world and in our sentimental impressions of the natural landscape as lush, tranquil, and beautiful.

A second example is more apropos to the discussion presented here, and of much more import. The gist of it is as follows: we can know nothing about a single particle in an otherwise empty universe because there is no context by which to measure size, mass, speed, or any other of its features. Such measurements only become possible when another particle appears, when comparisons can be made between two objects (distance apart, relative size, and so on). But note here that, instead of two entities existing in the universe, there are now three, namely, two particles plus the relationship between them; two material objects plus a non-material overlay of information which, if accessed, could provide a window into the properties of those objects which otherwise would not be available. With the creation of a second object, a third parameter, information, also comes into existence. This epiphenomenal parameter forms the ontological foundation for the ideational overlay familiar to everyday experience, and for the idiosyncratic interpretation of that experience which we would call 'impressions.'[4]

Being subjective, impressions can make no claim to infallibility. This is certainly true with respect to the relationship between the material and transcendent which, notwithstanding their underlying unity, is complex and multifarious, its various facets harmonious, or not, progressive, or not. As noted, there is now an inappropriate concordance between the ideational overlay and extant material circumstances in the form of the metanarrative which legitimizes the dysfunctional relationship between people and planet. The next chapter will take a deeper dive into this problematic duality, but I presage that discussion here by offering a preliminary, subjective impression of the material and transcendent domains in the context of the global environmental emergency, summarized under the headings of 'Climate Change' and 'God.'

Climate Change

I am not sanguine about climate change. Given that this problem presents an existential challenge to humanity – I say 'existential challenge' without qualification or hesitation – some justification of this opinion is called for.

Despite the fact that scientists have been issuing dire warnings to humanity for decades about the dangers of unchecked fossil fuel emissions, they often seem genuinely appalled, even frightened, when reporting on the rapid changes brought on by a warming world atmosphere. Whether it is glaciers retreating or Antarctic ice shelves calving or Greenland and permafrost melting or non-human species disappearing, or, in a nutshell, IPCC (Intergovernmental Panel on Climate Change) worst-case scenarios looming ever larger, the science community seems taken aback as if they did not expect any of these things to actually happen. Partly this is because Earth systems are reacting with surprising speed to heavy-handed human interventions. Partly it is because responsible scientists rightfully expect responsible policymakers to heed their warnings and to have already taken appropriate action, this hopeful expectation being underwritten by common sense and the logic of self-preservation. And partly it is simply because the technical analysis of the ramifications of a warming atmosphere lags far behind the curve of real-world disturbances forced by climate change. Such research is necessarily reactive and after-the-fact, a process further delayed by an inherent professional conservatism, by the need for caution and verification before 'going public' with findings in order to avoid publishing wrong conclusions which could fuel the fervour of the climate-denier crowd. Moreover, no scientist wants to be painted as alarmist so their personal impressions and, frequently, their deep concerns about the myriad implications of climate change, are not expressed except as the occasional disingenuous comment, "I was surprised."

That the planet is already experiencing serious turbulence caused by a much more energetic atmosphere is obvious and we the people, the apex predator and self-proclaimed hegemon of planet Earth, are still pumping up to 40 billion tons of carbon dioxide (CO_2) into the air every year. The expectation that, as stipulated by the Paris Agreement, we can and will rapidly reduce these global emissions is countermanded by the corporate power of the fossil fuel industry and by our collective addiction to cheap, plentiful, high-octane fuel, which explains why emissions are still rising (albeit more slowly), not falling. Renewable energy sources are displacing some small percentage of oil, gas, and coal use but for the most part new energy supplies are simply added to the overall inventory of energy required to meet the demands of a burgeoning human population on Earth.

Moreover, and contrary to popular perceptions, even a welcome but unlikely decline of CO_2 emissions to zero would not stabilize worsening atmospheric turbulence, though it would at least retard the process. First, there is a multi-decadal time lag between GHG emissions and their effects on the climate; in other words, there is already more warming coming down the pipeline. And second, all climate change models acknowledge that there is now too much GHG in the atmosphere. In order to predict the evolution of a stabilized climate all serious modelling requires that GHGs must be actively removed, bringing atmospheric concentrations down to, or as close as possible to, pre-industrial levels of some two centuries ago. While prototypes of 'negative emission technologies' (NETs) that can draw down CO_2 are being developed, none has even the remotest chance of being scaled up to global levels. It seems, frankly, preposterous to me that any such technology could scrub excess carbon out of some 1.2 trillion cubic kilometres of air without itself consuming gargantuan amounts of energy.[5] (An equally naïve possibility is 'bio-energy with carbon capture and storage,' a process so patently unworkable at scale that it does not deserve further discussion here.)

The alternative to a techno-fix is to rely on plant life (forests, primarily) and the oceans to absorb CO_2 'naturally.' Unfortunately, however, the world's two largest forest systems – the Amazon and the boreal forests of the northern hemisphere – are both under tremendous duress from both climate change and human interventions; they are both declining in vitality and size instead of thriving and expanding. And a warming ocean absorbs less, not more, CO_2. Current concentrations of GHGs, in other words, already the cause of tumultuous, expensive, and deadly atmospheric turbulence, cannot reasonably be reduced by any known method.

Additionally, most models of a changing climate do not include the warming effects of positive feedback cycles – most importantly, methane release from melting permafrost and undersea methane hydrates. There are many other such feedbacks that spur climate change; I will not list them here because they are already familiar to most observers. They are not, however, usually included in models because their effects are difficult to quantify. Not doing so, however, skirts the critical question of whether these cycles are now self-sustaining or not; that is, will they or will they not continue to exacerbate the climate change problem even if we can successfully and quickly reduce the forcing which kick-started them in the first place. I am dubious that the threat posed by these cycles can be attenuated.

And finally, the paleo-climate record in the history of Earth is not comforting. Our present situation is sometimes compared to an event called the Paleocene-Eocene Thermal Maximum (PETM) which occurred 55 million years ago, 10 million years after the extinction of dinosaurs. For reasons still under investigation, a massive injection of CO_2 into the atmosphere occurred over the course of several millennia, pushing the average global temperature up by some 5 to 8°C. It is of no reassurance whatever to know that the rate and quantity of emissions today, and the compressed time frame in which those emissions are happening, outperform the PETM by a factor of ten.[6]

In light of the intractability of these material issues and their potentially lethal implications, the best line of attack against this so-called 'wicked problem' should be mounted against its most vulnerable flank, namely, its metaphysical dimension.

God

Perhaps the most intriguing unanswered question now circulating in the philosophy of science is how to explain the fine-tuning of the universe, which is to say, why exactly so many fundamental parameters of physics have such precise values that, with only the tiniest difference in those values, neither stars nor galaxies nor life itself would be able to exist. While much ink has been spilled in pursuit of answers, only two are serious contenders: the multiverse theory, and, in one form or another, God. The first asserts that an infinite number of universes exist and of that infinite set one inevitably will comprise the parameters with which we are familiar. The second posits only one universe in which those finely tuned parameters have been set in place in accordance, perhaps, with a teleological impulse, to provide the physical basis for stability and propitious conditions for the evolution of life. Neither postulate can provide any empirical evidence to support its claims. The multiverse theory relies upon an extrapolation from known principles of cosmology and the manipulation of mathematical symbols. The God-theory calls upon such ineffables as the majesty of the universe and the intuition that our lives are meaningful in some larger sense, beyond mere care for our fellow creatures. Both theories are built upon interpretations of known phenomena; both are impressionistic.

I will present the argument in subsequent chapters that the God-theory is more compelling based on the evident fact that there is only one observed universe, all others being hypothetical; and that positing an infinite set of alternative universes is uneconomic and unnecessarily extravagant, allowing as it does for the actual occurrence, at some unknown place and time, of literally any and all possible events. Perhaps more importantly, I have identified the global environmental emergency as being the result of a violation of moral principles. If such principles are of the humanistic, situation-dependent kind, as they would be in a Godless universe, then their usefulness must be called to question, having evidently failed to delimit the profligate behaviour of our species. While profligacy might reasonably be blamed on things other than defective moral standards, it may also be reasonable to allow the possibility that self-made values built upon shifting foundations provide inadequate guardrails against our less desirable proclivities, and uninspiring incentives to seek a better way. Moreover, the violation of these free-floating principles has not prompted a re-evaluation of those questions which should accompany existential danger, questions such as why are we here, what are we doing, where are we going, and what, if anything, are we trying to accomplish? It may be that a move away from anthropocentric, socially constructed, and relative ethics to, instead, values drawn from a deeper well, values which carry with them the absolute imprimatur of supra-human force, might be in order.

Before advancing this argument, I acknowledge that there is what appears to be an irreconcilable problem at this intersection of the material and the ideational, that is, at the intersection of climate change and godhood. It seems to me that if the God-theory has any merit, and if in some sense humanity has been presented with the challenge of mediating between right and wrong, between good and bad – indeed, between life and death – then I would expect in fairness a level playing field on which these contests might occur. But ... if I were a businessman running a factory in England at the start of the Industrial Revolution, and if a local scientist introduced me to a black rock that burned, releasing copious amounts of energy; and if I could harness that energy to run my factory faster and more efficiently, then I would do so immediately and enthusiastically, powering my new machinery with coal. And soon gas and oil came along too and cheap, plentiful energy very quickly changed the industrial landscape of the entire world, helping to usher in the material bounty of what we now call modernity. It would be some 200 years before it was discovered that the massive combustion of fossil fuels could change Earth's climate in what we now understand to be an extremely destabilizing and dangerous way, but by then it was, for all intents and purposes, too late to abandon this richly abundant energy regime, too late to change our industrial trajectory in time to ensure a safe journey to the future. We had swallowed the hook too deeply. We have, arguably, injured ourselves fatally through no fault of our own. We did what anyone would have done under the circumstances, and with the information available at the time. It seems plausible to conclude from this historical narrative that failure was inevitable, that the playing field was never level. I have no answer for this puzzle.

That said, however, I have the stubborn impression that a better understanding of our present dilemma can best be found in a reconciliation of the transcendent domain where the self, the universe, and what may be called 'absolute values' are joined. I pursue this imaginary beginning in the second half of the next chapter with *The Transcendent World*, and from there through the remainder of this book. The first half of the next chapter, however, will begin by securing our understanding of *The Material World*.

Notes

1 This brief summary is drawn from https://aeon.co/essays/monist-phi losophy-and-quantum-physics-agree-that-all-is-one.
2 Entanglement in particular is now an integral feature of quantum computing.
3 For a novel exegesis of this position, see Wendt (2015).
4 It is also the root source of mathematics, which I discuss in Chapter 7.
5 One wag (David Orr) has called geo-engineering 'salvation by gadgetry.'
6 For more on this, see Hansen (2009), especially Chapter 8.

References

Hansen, James. 2009. Storms of my Grandchildren. New York: Bloomsbury USA

Rees, William E. 2023. "The Human Eco-Predicament: Overshoot and the Population Conundrum." *Vienna Yearbook of Population Research,* 21. https://doi.org/10.1553/p-eznb-ekgc

Wendt, Alexander. 2015. *Quantum Mind and Social Science: Unifying Physical and Social Ontology*. Cambridge: Cambridge University Press.

7 The Material and Transcendent Worlds

The Material World

'Relationalism' is an ontological theory that draws on the notion of dualism. This theory suggests that reality can only be made known to us by way of the relationships which occur between two or more entities. Position, size, mass, velocity, forces exerted or felt, even time itself are all without meaning for a single bit of matter in an otherwise empty universe simply because, according to relationalism, there is no context, no frame of reference against which to measure these quantities. Cosmologist/philosopher Lee Smolin puts it this way:

> [R]elationalism claims that the quantities physics can measure and describe all concern relationships and interactions. When we ask about the *essence* of matter, or of the world, we are asking what it is intrinsically ... the relationist stance is that there's nothing real in the world apart from those properties defined by relationships and interactions.
>
> (Smolin 2013, 266–7, original italics)

Primarily a mathematical physicist, Smolin maintains that knowledge must take the form of testable propositions, apart from which 'there's nothing real in the world.' The public verifiability of such propositions depends entirely upon empirical observation and the scientific method; metaphysical interpretations of reality are, on this view, particular, idiosyncratic, and decidedly unscientific, albeit not without personal value. This position, however, puts Smolin in his philosophical persona on the horns of a dilemma, which he

DOI: 10.4324/9781032647067-9

readily acknowledges. "But does it make sense," he wonders, "for two things to have a relation — to interact — if they are nothing intrinsically?" (Ibid., 267). Smolin finds no solace in the hope that mathematics, in its elegant and incisive ability to cut logically into the deepest mysteries of reality, might offer an answer. "If there's more to matter than relationships and interactions, it is beyond mathematics [because] relationships are exactly what mathematics expresses. Numbers have no intrinsic essence ... they are defined entirely by their place in a system of numbers" (Ibid., *Epilogue*, fn 10, 294).

Smolin's tussle with reality, briefly reviewed above, is hardly new in the philosophy of science. Immanuel Kant opined in the 18th century that "What may be the case with objects in themselves ... remains *entirely unknown to us. We are acquainted with nothing* except our way of perceiving them, which is peculiar to us" (Kant, quoted in Rohlf 2020, 22, italics added). And before Kant, in the 17th century, two heavyweights of the Enlightenment era, Isaac Newton and Gottfried Wilhelm Leibniz, fought to a draw about the nature of matter, space, and time, while René Descartes notoriously popularized the division of reality into two incommensurable parts, *res extensa* (matter) and *res cogitans* (thought), ultimately granting precedence to the latter.

Deeper in history and prior to the ascendance of practical reason during the Enlightenment, Thomas Aquinas in the 13th century suggested that observable objects were a combination of form and matter. A bronze bowl, for example, is made from base materials created at arm's length, as it were, by God but 'form' (such as the shape which makes it a bowl) is a subjective attribute expressive of God directly. And this in turn echoed scholars from antiquity such as Plato and Aristotle who expended much intellectual effort trying to come to grips with what it means to be real. Plato's influence on the debate has probably been the most long-lasting and influential, as will be seen. His dualistic notion of reality is nicely summarized by Philip Clayton:

> Plato found his solution in the doctrine of 'Forms.' What is ultimately real is the *eidos*: the idea of a thing. These ideas exist in a purely intellectual realm and serve as the pattern or exemplars after which all existing things are modeled. This object is a tree because it participates in the form of tree-ness,

and this is a just state because it participates in the form of justice.

(Clayton 2010, 40)

To be brief, the simple but perplexing fact is that we have no consensus, no verifiable idea what matter 'really' is, nor do we know, at bottom, what energy is either, except to say that it is interchangeable with matter by way of Albert Einstein's famous equation linking them together in what may be the world's best-known, elegant, but ultimately mysterious relationship. Einstein himself averred that the only way to bring any true understanding of reality into concordance with observation is by way of an "extra-logical (intuitive) procedure" (Einstein, quoted in Davies 1992, 80).

The relational point of view, especially with respect to observer dependency,[1] is gaining traction in the physics world, but in everyday life it is still just a sidebar to the materialist reductionism which has dominated practical thinking since the development of Newtonian mechanics in the 17th century. On this latter view, matter consists of small indivisible particles which, together with the four natural forces that act upon them, comprise the substance and behaviour of all objects in the world. 'Reductionism' simply means that analytically anything and everything can be reduced to these fundamental constituents and the laws that guide them; that all causal arrows point upward from below, and that all explanatory arrows point downward to this underlying aboriginal firmament.

Lacking in subtlety and obviously wrong, especially in light of Einstein's theories of special and general relativity and the emergence of quantum mechanics in the early 1900s, materialist reductionism nonetheless still dominates the public mindscape because, for all practical intents and purposes, it works. Newtonian mechanics inform the Machine Age. His equations linking force, mass, and motion are mathematically tractable and universally applicable. We build bridges, run factories, and put satellites into orbit according to those laws, and we can just as readily chart the movement of planets, stars, and galaxies with them. It is true that modern electronics and some important services such as GPS direction-finding are dependent on quantum phenomena, but by and large manipulating the 'real world' can be comfortably accomplished using Newton's straightforward but imperfect toolset.

Certainly, reality is clear enough to us – it is available to our senses without effort or any need for interpretation – and that has been sufficient for survival, adaptation (and, in modern terms, for prosperity) ever since our species began to walk upright. But we know now that it is radically incomplete. Our sense of vision, for example, upon which we rely for eighty per cent of sensory input, only responds to some ten per cent of the entire electromagnetic spectrum, the rest of which, ranging from very long radio waves to very short gamma waves, is invisible to us. This is analogous to our general perception of reality. We have a narrow sense of the physical (or the so-called classical) world through the lens of Newtonian mechanics which, helpful and useful though it is, excludes any understanding of what stuff is really made of. Probing the substance of reality more deeply with sophisticated quantum mathematics and experimentation still leaves an inaccessible residue of mystery and speculation. And this difficulty is multiplied by the fact that the cognitive tools we use to perceive and analyse the world are themselves no more accessible than physical reality, which is to say, we do not have any verifiable explanation for what consciousness is either. A reductionist ontology suggests that consciousness must inevitably be traceable to a 'brain state,' that is, to neurons firing in complex patterns, which in turn produces the phenomenon we recognize as a conscious 'mental state.' This conclusion is straightforward and understandable because much of what we perceive is simply the result of biological information processing and pattern recognition. These are things a computer can do and the human brain, on this view, is just an organic (albeit very complex) computer. This physicalist interpretation of consciousness still falls short, however, of explaining how neurons firing 'in here' can cause the perception of a world 'out there,' any more than it can explain why pain in one's knee is felt in that knee and not in the brain itself. This extension of consciousness beyond the brain and into the world presents a daunting problem for philosophers of mind. But the mind–body problem encounters an even more intractable obstacle in trying to explain the process by which the objective reality of neuroscience is transmuted into the subjective experience of living a life as 'me.' In other words, how can subjectivity emerge from objectivity? On this rock our understanding of the human mind founders: it is simply unknown how anything material could be conscious.[2]

Our sense of reality is bracketed on one side by the unknowability of matter and energy, and on the other by the ineffable quality of consciousness. The former somehow is presented to us as the classical world with which we are casually familiar; the latter gives rise to subjective experiences such as thoughts, emotions, memories, and a wide variety of sensations, all of which elude explanation even as we channel them into the construction of cultural phenomena such as music, literature, mathematics, logic, and spirituality. Only a very small percentage of the great mystery of the universe is available to us in knowable terms about which we may be categorically confident. And yet, this relative ignorance notwithstanding, we have achieved a truly remarkable portfolio of accomplishments; in fact, our intellectual prowess has produced much more than the material wonders of the modern age. Looking out and beyond, we have used our unique capacity for discovery on a grand scale to learn that we live on a small planet in the suburbs of an immense galaxy of stars, and that the Milky Way is just an average galaxy, one of some 200 billion others careering into the unfathomably distant reaches of the universe. We have even theorized the origin of the universe itself as having come into existence 13.7 billion years ago in what we now call the Big Bang. And from this marvellous ensemble of practical and esoteric knowledge we have discovered our own origin in starbursts, in atomic elements and the chemistry of life, and in the evolution of our species from the deep history of Earth.

The Transcendent World

Physical hegemony on Earth is one thing, but a masterful understanding of where we came from and in what stupendous surroundings we live takes us far beyond any mere material dominance and into a transcendent realm unreachable and unoccupied by any other species. This unique attribute is of particular relevance to the current environmental emergency because it allows us access to a non-material perspective, and thereby to the ontological domain in which narratives are born and evolve.

To be clear, I am using the term 'transcendent' here in its simplest possible meaning, to wit, not empiricism. It refers to knowledge gained beyond the senses, to the unobservable world of ideas and imagination (which, although completely open-ended, will be limited in the present context to that which comports with reason). Analogously, one may think of the 'transcendent world' as epiphenomenal or emergent, two terms I will use in subsequent discussions to enlarge the ambit of transcendency. Though I am sympathetic to the cognate philosophy of 'transcendentalism' espoused by Emerson, Thoreau, Kant, and various schools of Western or Eastern mysticism, idealism, Romanticism, and theology, that philosophy is not deployed here. The meaning of transcendent as used here is contained within the bounds of what is presented in this book.

What, then, can be said about this non-material, transcendent realm? Here dualism is pertinent once again, as described by Paul Davies:

> We come to know the world in two quite distinct ways. The first is by direct perception, the second by rational reasoning and higher intellectual functions ... Darwinian evolution has equipped us to know the world by direct perception ... but there is no obvious connection at all between this sort of sensorial knowledge and intellectual knowledge ... there is no selective advantage in our having brains able to incorporate quantum and relativistic systems in our mental model of the world ... The mystery is, why do we have this dual capability for knowing the world? ... After all, surviving 'in the jungle' doesn't require knowledge of the *laws* of nature, only of their manifestations ... Survival depends on an appreciation of how the world is, not of any hidden underlying order.
>
> (Ibid., 152–5, original italics)

And yet an understanding of the 'hidden underlying order' is precisely what we do have, notably with respect to the laws of physics, discovered through scientific experimentation and the logical manipulation of theoretical mathematics. Two questions are raised here: why do we have this access, and why does an underlying order exist in the first place?

The latter question has perplexed philosophers of science ever since physicists, stunned by their own remarkable success at understanding so much about the world (and far beyond), paused self-reflectively to ask 'Why is the universe intelligible?' and 'Why does science work?' Simplistic though these questions appear to be, they are not, as scientists began to realize that intelligibility was by no means a logical given; that the universe might just as easily have been chaotic and formless, indeed lifeless, except for the structure of the known laws of nature which, as far as we can tell, are absolutely invariant and pervasive across the entire known universe. The universe is ordered, not chaotic. But why is this so?

Here empirical scientists and interpretivists part company. The former either accept the world as it is without questioning the 'why' of it, or construct elaborate theories which purport the existence of a set of multiple, perhaps an infinite number of other, hidden universes in which all possible permutations of order and disorder occur (Greene 2011). Our universe is the way it is not because of preternatural circumstances but because, in an infinite set, one will inevitably look like ours. This point of view is vulnerable to criticism, however, as uneconomic, exaggerated, and incompatible with Occam's razor, an informal but useful principle which suggests that 'entities should not be multiplied beyond necessity' or, more commonly, 'the simplest solution is probably the correct one.' The transcendent point of view, which postulates the existence of a noumenal (perhaps divine) domain from which the laws of nature emanate, is more parsimonious. Its central claim is that our universe is unitary, unique, and entire unto itself, which agrees with what we perceive to be the case. This presents a very different context for the question of why the universe is ordered.

If the multiverse is dismissed as theoretically excessive, then the only way to explain the singular uniqueness of our universe is to illuminate some other reason why the laws of physics are what they are, and why so many constants of nature (from gravity to an electron's charge) are precisely tuned to build galaxies, stars, and planets, and to support life as we know it. This is tantamount to asking where those laws come from. One answer appeals to the Platonic realm of perfect Forms.

If the laws of physics came into existence at the time of the Big Bang, then they cannot themselves explain that event because they were not prior to it.[3] If they did exist prior to it, then where did they reside and what form did they take? Paul Davies suggests that pre-existing "transcendent laws of physics [may be] the modern counterpart of Plato's realm of perfect Forms which acted as blueprints for the construction of the fleeting shadow-world of our perceptions" (Ibid., 91–2). More succinctly, Werner

Heisenberg, author of the well-known uncertainty principle in quantum mechanics, avers that "Modern physics has definitely decided for Plato" (quoted in Ophuls 2011, 50).

Surprisingly, this cryptic point of view is not uncommon, partly because the field of mathematics in general is also troubled by a similar question. The laws of physics and the constants of nature are expressed numerically, but where exactly does the underlying logical structure of mathematics come from? Like the problem of order in the universe, this seemingly simplistic question also has no clear answer. One opinion called 'formalism' suggests that mathematics is a human invention, a numerical interpretation of order which otherwise has no substantive existence. It is a tool we manufactured to help us manipulate the material world. Another opinion argues that mathematics – number systems and the rules that govern them – is a whole unto itself;[4] it is discovered, not invented, because it pre-exists as an accessible but immaterial reality. Eminent mathematician and cosmologist Roger Penrose supports the latter perspective:

> Mathematical truth is something that goes beyond mere formalism ... It is as though human thought is ... being guided towards some eternal external truth — a truth which has a reality of its own, and which is revealed only partially to any one of us ... I imagine that whenever the mind perceives a mathematical idea it makes contact with Plato's world of mathematical concepts ... When one 'sees' a mathematical truth, one's consciousness breaks through into this world of ideas, and makes direct contact with it ... When mathematicians communicate, this is made possible by each one having *a direct route to truth* ... Since each can make contact with Plato's world directly, they can more readily communicate with each other ... communication is possible because each is directly in contact with the *same* eternally existing Platonic world.
>
> (Penrose 1989, 428, original italics)

Mathematics 'works' in my view because it is a true numerical translation or interpretation of ordered structures (of matter and energy) which inherently lend themselves to such interpretation, and in that sense I am a formalist. To put this differently, mathematics is invented but order is discovered.[5] The equation

E=mc² surely uses invented symbols, but those symbols represent a pre-existing relationship in nature which, by exploration, we have discovered. Mathematics is an epiphenomenal instantiation of the original underlying order of the universe; it is not a fully emergent entity with an existence of its own. If it were, it would have causal properties, specifically downward causation properties which would lend it the power to influence the ordered reality which it interprets. It does not have such power. In fact, sometimes the linkage between mathematics and reality is broken altogether.

There is no question that mathematics is logically structured, but some of its constructed edifices exceed the limits of material reality. One sees these over-extended structures in mathematical ideas and equations which, though perfectly logical in their own right, have no apparent connection to any physical entity. Or, they may purport the hypothetical existence of an unseen ordered structure or process which may or may not exist in which case mathematics can serve a useful predictive function, and, indeed, this function has achieved some notable successes.⁶ But, given the human capacity to manipulate epiphenomenal symbols with endless dexterity, constructed mathematical objects may simply be recreational, alluding to castles in the air which, though logically possible, are not really there. The multiverse theory may fall into this category.

Mathematics, then, provides a window into an ordered world, perhaps a Platonic world, but that does not help explain the source of that order. One obvious approach to the problem of explaining the unique features of our singular universe is to acknowledge the similarity between the Platonic world of Forms and the ineffable domain of a perfect God. It was Saint Augustine, an early Christian theologian and neo-Platonic philosopher, who made the relationship between the two explicit with the postulate that Forms were necessary components of the 'mind of God,' and that divine being was the actual foundation of the otherwise merely possible reality of those Forms. This understanding not only merges Plato's ontology with theology, it also opens a new window on the nature of consciousness as that which links our own sense of semantic awareness with the ultimate source of meaning in the world (and beyond), namely, God.

Of course, the issue raised here from an empirical point of view is that science is based on the concept of naturalism which allows no recourse to supernatural or transcendent forces or entities. Its objective is to explain *how* things happen by interrogating nature impersonally and experimentally. Theology, on the other hand, and cognate philosophies in general, are concerned with *why* things happen, relying on intuition and revelation to uncover reasons, not causes. Any successful merging of the two requires the mutual recognition that, first, science is not as concrete as its reputation advertises. Its observations and conclusions are inextricably theory-dependent (Bogen 2017) and much of what is studied remains fundamentally elusive, as previously discussed. And second, that the foundations of theology and morality are not as insubstantial as often portrayed; they are, like science, susceptible to reasoned argument and considered judgement. To deny this point, to draw a pejorative comparison between fact and opinion, would be to lionize science and dismiss as fanciful the work of philosophers, from Rawls and Kant to Aquinas, from Augustine and Socrates to Lao Tzu. Neither misrepresentation is supportable. An appropriate merger of science with informed intuition would consider the strengths and shortcomings of each and, in doing so, broaden our understanding of reality to include both how and why questions.

Notes

1 This refers to the fact that the outcome of quantum mechanical experiments will necessarily be influenced or determined by the characteristics of the observing apparatus.

2 *Panpsychism* supports the view that some form of 'proto-consciousness' is inherent to everything, right down to sub-atomic particles.

3 Note that this assumes the time-bound notion that cause precedes effect. Under the highly unusual initial conditions which obtained at the Big Bang, when time and space seem to behave differently, the cause–effect chronology may break down.

4 Theoretical physicist Max Tegmark argues that the universe is 'nothing but' mathematics. See Tegmark (2014).

5 It is important to note that, while mathematics is a powerful tool by which order can be discovered, it is not the only one. Any number of other tools – logic, intuition, utility, and spirituality to name a few – are also available.

6 For example, the planet Uranus and the positron, a sub-atomic particle, were both predicted mathematically prior to their discovery. Similarly,

the Euler beta function, a mathematical novelty, was later found to be directly relevant to string theory.

References

Bogen, James. 2017. "Theory and Observation in Science." In *The Stanford Encyclopedia of Philosophy*, Summer, edited by Edward N. Zalta. Metaphysics Research Lab, Stanford University. https://plato. stanford.edu/archives/sum2017/entries/science-theory-observation/

Clayton, Philip. 2010. "Unsolved Dilemmas: The Concept of Matter in the History of Philosophy and in Contemporary Physics." In *Information and the Nature of Reality: From Physics to Metaphysics*, edited by Paul Davies and Niels Henrik Gregersen, 38–62. New York: Cambridge University Press.

Davies, Paul. 1992. *The Mind of God: The Scientific Basis for a Rational World*. New York: Simon & Schuster.

Greene, Brian. 2011. *The Hidden Reality: Parallel Universes and the Deep Laws of the Cosmos*. New York: Alfred A. Knopf.

Ophuls, William. 2011. *Plato's Revenge: Politics in the Age of Ecology*. Cambridge, MA: The MIT Press.

Penrose, Roger. 1989. *The Emperor's New Mind: Concerning Computers, Minds and the Laws of Physics*. Oxford: Oxford University Press.

Rohlf, Michael. 2020. "Immanuel Kant." In *The Stanford Encyclopedia of Philosophy*, Spring, edited by Edward N. Zalta. Metaphysics Research Lab, Stanford University. https://plato.stanford.edu/archives/Spr2020/entries/Kant/

Smolin, Lee. 2013. *Time Reborn: From the Crisis in Physics to the Future of the Universe*. Toronto: Vintage Canada.

Tegmark, Max. 2014. *The Mathematical Universe: My Quest for the Ultimate Nature of Reality*. New York: Alfred A. Knopf.

8 The New Metanarrative
Some Ontological Considerations

There can be no soft landing for any transition to an ecologically secure and fertile future. Implicit in this assertion is a tension between reality and aspiration which can be expressed in terms of competing scenarios for the future, specifically, the business-as-usual scenario versus a new globally sustainable development trajectory. The former foreshadows climate chaos, widespread socio-ecological turbulence, and an existential shock to modern civilization. The latter anticipates a revolutionary shift in shared perceptions, attitudes, and values which will at least attenuate, and at best resolve, the accumulated effects of human profligacy over the last three centuries. Which one prevails will depend on the respective power of these competing stories. Either way, hard times lie ahead.

I have distanced myself in this book from any engagement with the brick-by-brick, bottom-up approach to sustainability because it inevitably hits the hard ceiling of a dysfunctional metanarrative which co-opts to its own purposes or otherwise renders ineffectual any material changes forthcoming. Progress, prosperity, and economic growth are deeply entrenched, popular and attractive ideas against which incremental adjustments have made a barely discernible impact. The reason for this is crystal clear, if embarrassingly simplistic. The incumbent metanarrative promises more of everything for everybody, imposing a redistributive burden on no one. For so long as the world economy grows without constraint or restriction, there is no need to share anything – with the poor who will soon be rich, with future generations who will benefit from the abundance of a technological legacy, or with other species who

DOI: 10.4324/9781032647067-10

will surely enjoy the commodious attitude engendered by wealth. Today's story of human life on Earth is the story of freedom from responsibility. It is a juvenile, amoral agenda which appeals to the callow mind of an immature society.

With regard to the evolution of a new metanarrative, I have suggested that new modes of analysis, including unorthodox methods and speculative hypotheses, may be useful. With this in mind, I argue in this chapter that the tension between ideals – whether conceived as Platonic templates, principles, or absolute values – and our worldly experience of them could be the critical dialectical driver of moral evolution, and therefore of the new metanarrative. On this view, aboriginal precepts of the Good, the True, and the Beautiful are mirrored imperfectly on Earth as morality, reason, and aesthetics. The inherent tension between these two domains, between the perfect and the imperfect, the necessary and the contingent, the timeless and the transitory, could provide the motive force for (r)evolutionary change.

Are Ideals Real?

Lest they disappear down the rabbit hole of irrelevancy, graduate students are often encouraged to keep in mind the purpose of their work by asking, 'what difference will this make?,' 'will this change anything?,' and 'does anybody care?' These questions are brutally summed up by supervisory faculty in their assessment of student research with the demand: so what?

It is a question that can be applied to almost anything because it immediately cuts to the core of the meaning of a situation – it is philosophically provocative, in other words. And nowhere has this query been more poignant than in a flippant remark once tossed off by an economist in *Business and Society Review*: suppose that, as a result of using up all the world's resources, human life did come to an end. So what?[1]

So what indeed. Unlikely though the prospect of self-termination might be, it does stimulate an interesting thought experiment which probes the meaning of such an unhappy eventuality. The null hypothesis (the position assumed to be true unless proven otherwise) no doubt implicitly endorsed by the economist just mentioned is ... so nothing. Extinction has no meaning. We did what we did because we could, and because we wanted to, and then it ended. We drove it until the wheels fell off. So what?

This is rude because it imputes to humanity an unflattering carelessness and a callow disregard for the future but, once again, so what? If human life did end, no one would be left to assign value to the loss of the human experience. Our absence simply would not matter. We would just be gone (the detritus of our civilization notwithstanding), a forgotten footnote in Earth's history.

The philosophical nettle here of course is that the null hypothesis would only be true if all value metrics – in fact the very concept of value, worth or merit – were social constructs, products of our imagination. If that were in fact the case, all such ethereal manifestations of human being would evaporate along with the physical ending of our species. On the other hand, however, if these things somehow exist independently of the human experience, then maybe our extinction (especially if self-induced) really would damage an unseen fabric, diminishing the goodness of the wider universe by foreclosing our participation in it. Our absence in that case would be genuinely misfortunate, perhaps wrong, even pathetic, according to some larger measure. This alternative hypothesis (the inverse of the null) posits that some important values, ideals, or principles may exist independently of human thought. If so, social constructions would retain their importance as worldly frames of reference but could now be understood as representations of less transitory, possibly universal, templates. This hypothesis matters – that is, this is relevant and could make a difference – because, if true, we would no longer be working for planetary sustainability for our own sake alone. What we do and how we do it could leave a lasting mark on the evolutionary trajectory of the universe itself.

Clearly, this allusion to native principles which may exist independently of humanity is controversial. The null hypothesis asserts that there are no absolute values on this planet, no inviolable truths or principles, and certainly no unconditional measure of morality. In a fruitless quest, deontologists, for example, call upon that which is intrinsically and timelessly right, that which finds expression in duty or obligation. It is exemplified by Kant's categorical imperative – Act as if the maxim of your action should become universal law – but it is not hard to find important exceptions to this requirement in a host of life's morally stressed situations.

Consequentialism, on the other hand, seeks retroactive guidance for rectitude in the results of conduct instead of the behaviour itself. Jeremey Bentham and John Stuart Mill are the progenitors of its best-known variant, utilitarianism, which seeks the 'greatest good for the greatest number.' But here too, proponents run afoul of practical calculations such as how to measure 'good' and how to balance costs and benefits over an indeterminate number of people, and an indeterminate period of time. It seems clear that, apart from the laws of physics, nothing is necessarily true. This bald assessment implies that all morally vexed situations are contingent affairs that could have been otherwise under different circumstances, and this in turn raises the possibility that all ethics are relative, that morality is situation dependent; that postmodernists might be right after all.

It may very well be true that there is no perfection 'on this planet' and that morally vexed situations are contingent, but this does not necessarily obviate the alternative hypothesis, the wider possibility that valuation templates – ideals, as it were, from which worldly notions of truth, morality, and aesthetics emerge – exist prior to our experience of them. To assert otherwise, to insist that these principles exist at the whim of the imagination, is to arrogate to ourselves authorship of all determinants of valuation. From an intuitive or revelatory point of view, this is disturbingly prideful, a wilful exaggeration of human exceptionalism.

Granting the existence of ideals opens the possibility of comparing them with observable behaviour. Allowing this, then, and regarding our worldly experience of them, it is abundantly and immediately clear that we have not done very well in the practice of reflected ideals. Morality is strained by war, by the ill-treatment of our children, and by the expropriation of the living space of Earth's other creatures. Reason is compromised by the destruction of resources vital to our survival despite clear lessons from science and history. And aesthetic sensibilities are dulled by rampant industrialism and the defacement of natural landscapes. This inappropriate behaviour is exaggerated by a collective sense of exceptionalism which we have taken to mean that we are independent of nature, of any kind of loftier reality, and that we are unaccountable for our actions. The hypothetical positing of supra-human precepts corrects these misapprehensions by providing a new context for evaluation – by moving us from the

fractured values inherent in postmodern (and post-truth) secularism to truer values drawn from a deeper well, and to a better
understanding of our moral predicament as a species in self-
induced peril.

The Emergent Status of Metanarrative

Because they are immaterial, stories (like ideas in general) can only
exist in the noumenal domain.[2] Whether that domain is a mental
construction with no meaningful extension beyond the individual,
or whether it can be shared among multiple people, or whether it is
joined with a more perfect (Platonic, perhaps) realm with which it
interacts are questions without definitive answers. I have suggested
that the last of these possibilities is the most useful in terms of
understanding the status and meaning of metanarrative and, more
importantly, how it might evolve into something better suited to
the objective of achieving planetary sustainability and a fruitful
future for human society on Earth.

Native principles, supra-human precepts, and interaction with
the Platonic realm are promising phrases that admit the possibility
of absolute values independent of human construction, and as
such they introduce a tension between the perfect and the imperfect, between the timeless and the merely temporal. This tension,
however, cannot be unambiguously resolved. Bridging the gap
between the Earthly realm and a perfect world is not possible,
though the latter may be approached asymptotically. There is no
such thing on this planet as infallibility, so one need not anticipate a modern version of the Ten Commandments or any other
professed version of the 'word of God' to define morality or, for
that matter, sustainability. The benefit we share in the acknowledgement of an unreachable foundation on which our values are built,
and from which they emanate, is that such knowledge transports
us beyond ourselves and invites us to participate in a more profoundly interactive performance. It makes real what would otherwise be nothing more than a pedestrian exercise in humanism. The
interesting question at play here is how one may approach perfection asymptotically. The following preliminary discussion takes up
this issue with reference to the phenomenon of emergence and, in
the closing section, agency.

Perhaps the earliest expression of the interactive unity of the individual with all else occurred in the ancient Hindu Upanishads. Swami Prabhavananda tells us, for example, that

> The little space within the heart is as great as the vast universe. The heavens and the earth are there, and the sun and the moon and the stars. Fire and lightning and winds are there, and all that now is and all that is not.
>
> (Upanishads quoted by Anonymous 2002)

This poetic rendering of the linkage between the individual and the cosmos finds some modern-day resonance in mysticism and meditation but, from a secular or academic point of view, the ontological status of such views is unclear at best. Along with religious dogma and revelation, today's reductionist socio-political mainstream tends to group such phenomena under the heading of 'ideas' which, whether individual, shared, or otherwise, are somehow lifted out of the material world and therefore not susceptible to scientific inquiry. Notable exceptions to this trend are sociologists such as Émile Durkheim who take the view that ideas are no less natural than material reality; that they exist as 'social facts' such as liturgical practices, legal codes, money, property, marriage, and so forth, the real presence of which can be inferred from objective indicators of their social expression; and this grounding allowed him to remain within a conventional positivist framework. This obviously is a much more prosaic rendering of the transcendent domain than that offered by the Upanishads, limited as it is to a humanist world, to ideas shared only among people and nothing beyond. For Durkheim, social consciousness is not revelatory, nor is it a reified entity with an independent and unknowable existence, but merely a set of ideas common to many individuals. Philosopher John Searle agrees, arguing that collective intentionality does not require "the idea that there exists some Hegelian world spirit, a collective consciousness, or something equally implausible ... because intentionality remains in individual heads, existing in the form 'we intend' and 'I intend only as part of our intending'" (Searle 1995, 25–6). Any number of alternate interpretations of the status and reach of a shared consciousness, and therefore of a dominant metanarrative, might be found. Jean-Jacques Rousseau's bipartite expression of a 'will of all' versus a 'general will' offers a useful way to summarize these competing viewpoints.

It is a common fact of life in modern times that we are much more comfortable with the notion of individual sets of experiences and preferences which may be aggregated into sum totals, but which do not together constitute an independent entity with unique characteristics. Rousseau called this sum total the 'will of all.' Interestingly, it is precisely this will of all, this aggregation of personal experiences and preferences, now called utility functions, which provide the rationale and raw data for our consumer-oriented economies. The opposing point of view, however, stipulates that the characteristics of a human group cannot be derived simply by aggregating the attributes of individual members. The 'personality' of a group is generated by the pattern of organization and dynamic interdependencies among members of the group. That personality is an emergent phenomenon with an independent ontological status and unique causal powers. In support of the notion of a shared consciousness and in contrast to the will of all, Rousseau posited the existence of a 'general will,' a unified expression of social concern about the common good, a collective impulse to achieve the best interests of society as a whole. Rousseau's famous social contract was intended to create the political space in which this impulse might flourish (Rousseau [1762] 1987).

The metanarrative of Progress and Prosperity might be described as an expression of this general will – but its ostensible purpose to realize 'the best interests of society' has missed the mark because its widespread influence rests largely on a radically oversimplified mantra (more of everything for everyone) as previously discussed. A more mature, sophisticated story could not build upon such a crude aphorism, but instead would have to give voice to the richness of the whole human experience while still embracing the particularities of cultural diversity, economic inequality, and all sorts of other important differences between peoples and civilizations. But we are a long way from that. Instead of an informed and progressive general will we are more likely to see in the global polity a continuation of simplistic expectations and crass, self-serving behaviour along with indiscriminate waves of fear and panic, anger and excitement, guilt and sympathy spreading fast across large sectors of society, sweeping up individual sentiments in the psychological volatility of the undisciplined 'mind' of large groups of people. We are more likely to agree with Carl Jung who said, "The masses always incline to herd behaviour, hence they are easily stampeded; and to mob hysteria, hence their witless brutality and emotionalism" (Jung 1985, 6). And, in a similar vein, William Ophuls has said that "The greatest weapon of mass destruction on the planet is the collective human ego" (Ophuls 2011, 72). These harsh observations seem closer to reality than the putative benevolence of an unformed general will.

Nonetheless, emergent properties carry the indispensable features of innovation, adaptability, and evolution and therein lies the possibility of positive and productive change. This has not happened yet, but complex adaptive systems and their peculiar ability to spawn such properties do sensitize us to the possibility that the human population may in fact be capable of generating such a phenomenon as a shared consciousness, however crude or unformed such a thing might be now. The impression offered here is that this is indeed the case: a nascent collective consciousness does exist and, like Rousseau's general will, it could under the right circumstances take the form of a common intention to ensure the vitality and stability of the socio-ecological complex. If this were to occur, our future would be greatly benefited.

To elaborate briefly on this, emergent properties are extra, unanticipated features which emanate from the synergistic interactions of a complex system's component parts. The familiar expression 'The whole is greater than the sum of its parts' loosely captures this phenomenon. The ability to perceive and engage with intangibles such as ideas, values, beliefs, and worldviews is an emergent property of the individual human body and brain, and the supposition at play here is that a similar property can be ascribed to the human population as a whole.

Emergent properties become manifest as one moves (synchronically) up the ladder of complexity from atoms to the cosmos,[3] or as one moves (diachronically) through time with respect to, for example, biological evolutionary history. Novelty and wholeness are clearly visible features of unique phenomena which occur at each new level of organization, at each step in the progression from physics to chemistry to biology to psychology to sociology and beyond. Such features are not reducible to components, or to previous levels or stages. Moreover, all are affected by and have the capacity to affect the behaviour of adjacent structures. The larger picture, then, presents a series of nested systems, each unique unto itself, each involved in a dynamic whole/part relationship with its neighbours according to which that duality is mutually constituted, and according to which the behaviour of parts is influenced by supervening emergent structures.

So-called 'downward causation' does not compel activity by adding new forces of nature to the four already known; rather, emergent structures such as a putative collective consciousness (and the metanarrative it instantiates) make constituent parts behave in ways they otherwise would not by guiding and encouraging, by imposing determinative influences within the bounds of physical law. Contextual effects, constraints, and novel patterns of organization are the result. Just as the material environment affects biological development, so society affects the personality and preferences of individual people, and a shared worldview affects the evolutionary development of human society on Earth.

Relationalism Redux and Agency

According to the descriptive ontology discussed above, the individual person is implicated with the cosmos through a series of complex structures and emergent features, all dynamically linked in an evolving whole. More broadly still, this continuum stretches not only from person to cosmos, but from the very primordial stuff of existence to, from a Platonic point of view, "a sentient universe charged with moral meaning" (Ophuls 2011, 21), that is, to the foundations of moral order implicit in the natural world. There is a distinct similarity between this point of view and relationalism because, in its fullest expression, the latter avers that all things are interconnected, interrelated, and interdependent, and that people, being an integral part of this matrix, are all co-constituted and co-dependent. On this view, the world is an assemblage, an entangled network of relations.

In a broad sense relationalism stands opposed to Western notions of individualism and individual agency because the self no longer has the independent authority of an unattached person. Instead, intentional action is thought of in terms of 'distributed agency.' In the end this sophisticated and complex philosophy, which has deep roots and considerable appeal, does allow some room for novel action, albeit informed, enabled, and constrained by inescapable contextual parameters. For me, however, burying individual agency in this tangle of relationships limits its usefulness, its readiness for action, one may say. A more kinetic approach to individual agency is necessary to respond to the environmental danger now afoot, and I find this in Hannah Arendt's more parsimonious treatment of agency (see Chapter 9) which calls upon 'natality' (individual uniqueness) and 'plurality' (the context of that uniqueness). These features are comparable with those of relationalism but they single out individuality more sharply and, in effect, grant more power to it. And herein lies the key to agency with respect to the evolution of a new metanarrative. Notwithstanding the level of abstraction inherent to the idea of a metanarrative, and even though my focus in this book has been on human society writ large, there can be no doubt at day's end that true agency begins with, and emanates from, the individual person. That person, an independent but responsible/responsive agent, can reasonably be construed as being at one, so to speak, with an implicate universe. This dynamic relationship joins individual consciousness with the shared ideational domain, and brings it to proximity with the supra-human virtues which also inhabit that space.

Notes

1 This anecdote was shared by Gare (1995, 12).
2 'Noumenal' refers to entities not perceptible by human senses. It stands in contrast to the phenomenal world of sense perception.
3 Arguably, the universe as a whole is a complex adaptive system.

References

Anonymous. [2002]. *The Upanishads: Breath from the Eternal.* Translated by Swami Prabhavananda and Frederick Manchester. New York: Signet Classic Books.

Gare, Arran, E. 1995. *Postmodernism and the Environmental Crisis.* New York: Routledge.

Jung, Carl G. 1985. *The Practice of Psychotherapy.* Translated by Gerhard Adler and R. F. C Hull. Princeton, NJ: Princeton University Press.

Ophuls, William. 2011. *Plato's Revenge: Politics in the Age of Ecology.* Cambridge, MA: The MIT Press.

Rousseau, Jean-Jacques. [1762] 1987. "On the Social Contract." In *Jean-Jacques Rousseau: The Basic Political Writings*, translated and edited by Donald A. Cress, 141–227. Cambridge, MA: Hackett Publishing Company.
Searle, John. 1995. *The Construction of Social Reality*. New York: Free Press.

9   Agency and the Evolution of Metanarrative

In his well-regarded book *Moral Man and Immoral Society* (Niebuhr [1932] 1960) American political theorist and theologian Reinhold Niebuhr explains that individual people are reasonable and morally sensible, but that reason and morality diminish from individual to group, and diminish even further as group size grows larger.[1] Values important to individuals lose their salience in the context of group behaviour, so larger groups are less responsive to reflexive evaluation of what they are doing. This means that reasoned, morally sound supervision of our collective behaviour is weakest when the group is biggest, and the biggest group of course is the human population as a whole. The result of this is that human society on Earth is easily tantalized by simplistic ideas and momentary impulses, easily stampeded into unreflective, emotion-driven behaviour which lacks subtlety, sophistication, and direction.

But change is possible. Unlike machines, complex adaptive systems and their associated emergent properties evolve over time. Thus, it may be true that, like the individual human mind, humanity's collective consciousness can evolve through developmental stages, from a rudimentary awareness of the world around, through a kind of adolescent excitement and vulnerability, and finally to a more mature and stable form characterized by empathic care and a sense of adult responsibility. This latter stage of consciousness – the stage at which something like Rousseau's general will, or a new metanarrative, might emerge – would be well suited to serve the public interest by constraining or enabling the macro-behaviour of human society as circumstances require. It seems

DOI: 10.4324/9781032647067-11

clear enough, however, that we are currently at the middle stage of development – excited, vulnerable, and easily diverted from serious consideration of the long-term health and vitality of human society on Earth.

This has clear implications for the instigation of change. If humanity's collective consciousness is now at an intermediate phase of development and maturation; if, as such, that consciousness is susceptible to Panglossian ideas; and if, consequently, human society has embraced the ecologically unsustainable notion of unbounded progress and prosperity, then a way must be found to move the process of social evolution forward, to countermand this juvenile agenda and to put in place a more responsible plan. This means subsuming the incumbent metanarrative within one which serves the better aspirations of people, and encourages the pursuit of those aspirations with a certain sense of ecological modesty.

One may speculate that this process of maturation is natural, organic – it occurs as the result of accumulated experience and the gradually improving ability to assess the value of that experience. Such a pleasant journey, however, seems out of the question. Humanity is rushing headlong into tremendous socio-ecological turbulence which may or may not be survivable. This is not an avoidable fiction. Biophysical indicators are all pointing in the wrong direction, powerful forces are arrayed against change, ethics are a devalued commodity. And, the creation of a new metanarrative usually requires a decades-long process of social learning. William Rees is not sanguine about the chances of success:

> Do we have decades and the resources to pull this off, particularly since the traditional means for social learning are retreating from ... the legitimization of 'fake news' [and] the self-serving echo-chambers of social media? ... [A] new global metanarrative is, in fact, taking hold but one rooted in tribalism, individualism, competition, social discord, science denial and populist discontent ... I cannot imagine a real-world process whereby the great fractious heterogeneous mass called 'we' can create a new world order of progress and prosperity shaped by a stable, ethically informed socio-ecological metanarrative. Stability is no more; ethics are for sissies; there is no socio-ecological compromise.[2]

A first response to this bleak assessment would hearken back to the point made that metanarrative evolution is not necessarily reliant on the coercive tools usually available to powerful political and corporate interests, such as money, material resources, and media ownership. People also respond to the creative humanities in the form of stories, metaphors, and allegories which facilitate a morally sensitized emotional engagement. Ideas, in other words, especially those disseminated by omnipresent social media platforms, can inspire rapid, even revolutionary change, sometimes virtually overnight.

Second, in a world dominated by self-interest and materialism, and by the political exercise of the crude philosophy that 'might is right,' one could remark that there remains in human society, and in the hearts and minds of individual people, a lingering notion that ethics do matter and that the aspiration to a morally grounded society is a worthy objective. The challenge at hand is to operationalize this aspiration and thereby to impel and re-direct the co-evolutionary process which binds us to the universe.

And third, we need not wait – indeed we dare not wait – for 'social forces' to come together in an ecologically auspicious moment. As noted earlier, change must inevitably begin with the individual. Personal agency is a radically underestimated resource. The potential to enact transformational change is neatly captured by Hannah Arendt in her book *The Human Condition*. That agency begins from her observation – profound in its simplicity – that the birth of every child represents a new beginning by bringing novelty into the world. She calls this 'natality.'

> It is the nature of beginning that something new is started which cannot be expected from whatever may have happened before. This character of startling unexpectedness is inherent in all beginnings ... The fact that man (sic) is capable of action means that the unexpected can be expected from him, that he is able to perform what is *infinitely improbable*. And this again is possible only because each man is unique, so that with each birth something uniquely new comes into the world.
>
> (Arendt 1958, 177–8, italics added)

Arendt is not insensitive to the context in which each person acts, a context she refers to as 'plurality.' This term expresses both equality, inasmuch as we are all of the same species inhabiting one planet, and distinction (novelty) because "nobody is ever the same as anyone else who ever lived, or will live" (Ibid., 7–8). No two people are interchangeable, no two have the same relationship with the living world. Uniqueness and sameness (the individual and the whole) exist simultaneously and in fact agency requires their union because

> [bringing the unexpected into the world] …is not something that can be done in isolation from others, that is, independently of the presence of a plurality of actors who from their different perspectives can judge the quality of what is being enacted. In this respect action needs plurality in the same way that performance artists need an audience; without the presence and acknowledgement of others, action would cease to be a meaningful activity. Action … can only exist in a context defined by plurality.
>
> (Arendt, quoted in Passerin d'Entreves 2019, 17)

Writing in a similar vein, philosopher Arran Gare captures the dynamic element of the relationship between embodied subjects and contextual parameters in his quest for a new cosmology. Individual action, he writes, should

> neither atomize the world nor dissolve each part into the totality … [people should] experience themselves as processes of becoming, actively participating in the becoming of the world … by identifying and comprehending the dynamics of and interrelationships between a multiplicity of semi-autonomous processes.
>
> (Gare 1995, 155–7)

The successful operationalization of these comments and observations – which is to say, the realization of our own quest for a new cosmology – will surely begin from, and draw upon, these notions of natality, plurality, and participation in interwoven processes of becoming from which, together, creativity will emerge. An early burst of human creativity, well-intentioned but morally unsound, has already spawned the Anthropocene epoch and the serious challenges it portends, so the new story we tell will unavoidably feature self-inflicted danger. But it will also offer the opportunity to re-imagine the exceptionalism that ennobles the human animal, and to rejoin our unique endowments to the planet that gave us life. The Chinese trope about the coincidence of danger and opportunity may be appropriate here, so long as opportunity is defined with regard to the moral transfiguration of the iniquity which brought us to this dangerous juncture, and not in terms of returning to the mythical cornucopia of Eden. Our first challenge as adults will be to face the consequences of past behaviour with courage, and even grace. Only having done so will we be able to fully engage the possibility of exploring at our leisure the true and unlimited potential of all that we are, and all that we can be.

Individual Agency

Much is expected from the individual person according to insights offered by Hannah Arendt and Arran Gare, justly so because individual people are the primary loci of change, but the successful operationalization of these comments and observations – which is to say, the realization of our quest for a new cosmology – remains a formidable problem. It seems presumptuous, perhaps even pointless, to call upon people everywhere to live life like a story, to 'participate in the becoming of the world,' to personally acknowledge the existential reality that human society is at an intermediate (adolescent) stage of development and must somehow find a new path to maturity. And yet the bottom-up, citizen-oriented quest for change is still the *modus operandi* of contemporary environmental activism. Even when personal agency is embedded in a social context – Arendt's plurality – it tends to fall short of its objective because, without aggressively tackling the transformation of the world's ideational superstructure, it remains essentially pragmatic. It amounts to people rising against a status quo they deem unacceptable, people rising against the holders of power and the institutions which normalize and legitimize their superior position in society. This battle is being undertaken in earnest, but only by a minority, and with only modest success. Arguably this blunt force for change needs a sharp point without which the veneer of normalcy which protects the status quo like so much chainmail will remain impenetrable. The alternative to a mass citizen-led movement is vanguard leadership. Though often disparaged as prone to abuse and corruption, I consider it below.

Arendt offers the observation that all people are the same (members of a single species on an indivisible planet), yet each of us is unique. The first of these fits nicely with the philosophical tradition which stipulates that all people are of equal worth, and bear equal dignity. The second accords with the practical observation that our different personalities and predilections take each of us down different roads in life. The inexorable result of the interplay of these differences, however, is that human societies, almost exclusively, settle into hierarchical structures. This latter observation opens the door to what may be the main and most contentious theme of political sociology, namely, social stratification.

The natural tendency to hierarchy is typically represented by the term 'class.' Max Weber saw a multiplicity of class cleavages in his development of a pluralistic, multidimensional model of social stratification. In contrast, Marxists retained the traditional conception of class as a dichotomy between pairs of antagonistic groups such as master-slave, feudal lords and serfs, or capitalists and workers in the modern era. Though more parsimonious, the Marxian model shared with Weber's the belief that class is just a byword for patterns of domination and subordination against which impulses to egalitarianism are perennially but ineffectually opposed. On this view, class typically carries the pejorative cast of oppression and unfairness.

Even given the inevitability of stratification, however, it does not necessarily follow that brute domination is an unavoidable feature of leadership. A functionalist interpretation, for example, would suggest that stratification occurs naturally as people follow their own personal interests and abilities. "Although many can cook, few can do theoretical physics," as William Ophuls expresses this point (Ophuls 2011, 115). He goes on to suggest that "[S]ome form of aristocracy is essential for a healthy democracy ... only a genuine elite can supply the discipline, prudence and forethought necessary to check the destructive tendencies of crowds" (Ibid., 150).

Plato's *Republic* (Book III) offers a similar version of 'form follows function' with his tripartite metaphorical depiction of good governance as the division of people into gold (systemic), silver (managerial but not systemic), and bronze (practical) types which work cooperatively together in the construction of the ideal polity.[3] Those with a temperament suited to leadership can be indoctrinated into their roles by way of *paideia* (elaborated in Aristotle's *The Politics*, Book VII), a process of rearing and education which includes diversified training in the liberal arts, in science, and in physical exercise – in other words, a full regime of mind and body improvement and vitality. Significantly, Plato's interpretation of a natural aristocracy also includes the controversial notion of philosophy for the few but the promulgation of a 'noble lie' for the many. By this 'lie' he means an optimistic, rose-tinted story about present conditions, and about the future, in order to bolster the psychic stability and well-being of the masses as they go about their quotidian routines. Reinhold Niebuhr also speculates that a fictional story may be required to calm the public waters. People, he suggests,

> will have to believe rather more firmly in the justice and in the probable triumph of their cause, than any impartial science would give them the right to believe, if they are to have enough energy to contest the power of the strong.
>
> (Niebuhr, xv–xvi)

In the modern political context, Thomas Jefferson, James Madison, John Locke, and Jean-Jacques Rousseau were all, in their way, proponents of elite political leadership. Madison, for example, said

> The aim of every political constitution is, or ought to be, first to obtain for rulers men (sic) who possess most wisdom to discern, and most virtue to pursue, the common good of the society; and in the next place, to take the most effectual precautions for keeping them virtuous whilst they continue to hold their public trust.
>
> (Hamilton et al. 2010, 165)

Many find these views objectionable and quite possibly dangerous, but they need not be. Leadership is not just about the exercise of preponderant power by a dominant class; it may equally emphasize the consensual nature of hierarchical social arrangements. Hegemony as understood by Antonio Gramsci, for example, certainly includes a profound measure of social and moral authority but likewise depends on the alliance of ordinary people who have been won over by concession and compromise. The resultant leadership paradigm is built upon foundations which are cultural, intellectual, and, most importantly, ethical. This discussion brings to mind the phrase 'noblesse oblige' which refers to the fact that privilege also incurs responsibility, the objectives of which can readily be understood as the many social, economic, political, and ecological necessities of sustainable development listed in Part I.

The top-down strategy for change, discussed above in the guise of elite leadership, also includes the hopeful prospect of a newly evolving metanarrative. This latter objective, however, has not fared well, so far making no impression at all on our collective psyche. A substantial constituency of people around the world is now in the process of trying to create a new metanarrative by drawing on extant documents such as the Earth Charter as well as a plethora of cultural narratives which speak to various aspects of sustainability. By extracting expressions of value from each of these, a universal set of core ideas, sometimes called memes, might be put together which would resonate strongly enough with people everywhere to overwhelm the opposing set of ideas which now constitute the structural elements of the incumbent metanarrative. A multitude of such memes has been identified, including virtues such as holism, interdependence, stewardship, well-being, equity, sufficiency, respect, ecocentrism, inclusiveness, dignity, and humility. The hopeful expectation is that the widespread dissemination of these memes in the form of a coherent narrative will generate the transformative change required to achieve the necessary trajectory correction. But many of these words are obscure, almost sanctimonious and without immediate relevance; they do not resonate in the public mind as clearly or as personally as does the siren song of Progress and Prosperity, and they are certainly not heard in the halls of power, even with environmentally literate, progressive governments such as those which typically represent the Canadian polity. The putative benefits of Plato's 'noble lie' notwithstanding, it may very well be futile to project a bucolic fantasy into a future which will inevitably be more unstable, less accommodating, and more dangerous than is optimistically envisioned. Any new metanarrative, to be relevant now and to the future unfolding before us, must be tougher, cognizant of the wrongfulness that brought us here and more vividly expressive of the moral maturity without which our potential will remain unfulfilled.

Optimism and Hope

An alternative strategy to the dissemination of platitudinous memes is to insinuate into the public domain, with whatever gravitas can be mustered, a harder, more pessimistic viewpoint paraphrased as 'brace for impact.' An evidence-based, life-and-death polemic might cut through the bustle of everyday life, giving reflexive pause to people everywhere, laying the groundwork for a new story about the human experience on Earth. But, evidently, this will not work. Urgent warnings and graphic scenarios highlighted by respected professionals from multiple disciplines have had no lasting effect. Also, deniers, profiteers, and self-serving political actors work actively to portray such scenarios as alarmist and irresponsible, therefore lacking credibility. And, unfortunately, brutal honesty can cause negative responses such as despair, panic, or violence, rather than constructive, cooperative action.

The counterpoint to frank honesty is a kind of steely optimism. It is common knowledge, after all, that optimism is enabling. It may misrepresent the real world if overblown, but it is hard to argue with the observation that pessimism can be self-defeating while optimism keeps enthusiasm alive in the face of setbacks. And, it is free – free in the sense that it does not arise from, nor is it dependent upon, any empirical referent. No matter what the brute facts of the world may be, we retain the capacity to maintain a positive attitude.

Having said this, it is sobering to see that for many, hope is not far removed from cynicism. According to at least one survey (GlobeScan 2012, 20), more than seventy-five per cent of sustainability experts around the world agree that planetary sustainability can only be achieved with a stick, not a carrot. They believe that concerted political action to avert ecological catastrophe will not occur until a truly awful, unmistakable event takes place – a massive drought-driven famine of historic proportion, say, or a shockingly violent tornado swarm, or perhaps an abrupt change in oceanic currents. Only then will we be spurred to action in a last desperate drive for survival. This is a morbid kind of hope to be sure, but it does seem typical of human behaviour. The point is that facile thinking about a rosy future cannot be squared with ever-mounting planetary stresses and the constant breaching of worst-case scenarios. A more honest approach is to take owner-ship of the consequences of irresponsible behaviour, to confront the damaged relationship between people and planet as a real-life spectre instead of dismissing dire scenarios as incompatible with a positive attitude, as counterproductive to progress.

I have conflated optimism and hope in the foregoing, but it is commonly recognized that they are not the same thing. Optimism is simply a confident attitude that a desirable outcome will be achieved regardless of any empirical referent. Hope also looks forward to achieving something desirable, but it is tempered by an assessment of the measure of possibility. We can rationally believe that the chances of success are low, we can be pessimistic, expecting the worst outcome, but persist nonetheless because the outcome we seek is valuable, and may even be possible. Unlike optimism, hope is vulnerable to disappointment and failure, but its exercise helps us to ascertain what is valuable, and what price is worth paying.[4]

Vaclav Havel agrees, but puts a somewhat different spin on the meaning of hope. He suggests that "Hope is not the same thing as optimism. It is not the conviction that something will turn out well, but the certainty that something makes sense, regardless of how it turns out" (Havel 1986). This point of view does not turn on the possibility of success, but only on an achieved understanding of what is happening so, at a minimum, we are not confounded by life or confused by the outcome of our actions; and this opens a new way of looking at our planetary predicament. If we can 'make sense' of what we are doing to ourselves and to the planet, if we can find comprehensible reasons for having taken the path we are on now, then a new conceptual frame may come into view which can drive a more incisive and effective kind of agency for change. A hopeful posture in the Havelian sense may help shift the analyt-ical terrain from how questions (How can we solve the problem of planetary sustainability?) to why questions (Why do we face an existential predicament?), adding a different kind of impetus to the quest for transformative change. Far from being an exercise in epistemological futility, this shift could encourage an exploration of deeper currents and more profound truths which may lead to new understandings, even to new modalities of global governance and to the breakthrough we clearly need on the road to planetary sustainability. The search for meaning is an underdeveloped aspect of the modern human experience. A new effort to make sense of a dire situation may very well hold useful surprises for us, revelations or even an epiphany, a shared awakening to the meaning of the life-and-death struggle in which we are now engaged.

A top-down force for change is a necessary complement to bottom-up public pressure. Each is (or should be) reciprocally constituted by the other, each gains its legitimacy and cogency from the other, and both together are necessary components of the thrust to a new reality. This possibility is most often discussed in a political context but, with respect to the evolution of a new metanarrative, a different context, and different actors, is necessary. If the challenge at hand is to expedite the moral evolution of modern human society in order to render the global environmental crisis manageable – or at least to arm ourselves with the courage and foresight needed to cope with radical disruption without turning violently on ourselves – then a new kind of leadership is required. Though one may hope for timely political intervention, I expect this vanguard to arise instead from practitioners and theorists of the humanities. Where are our modern philosophers? Where are our poets and storytellers? Where are our moral leaders? I am not referring here to the narrow, and usually exclusive, religiosity of institutionalized belief systems, nor to the technical wizardry of sophists whose words are incomprehensible, and therefore meaningless, outside the academy. I am referring instead to a new generation of thinkers and doers, people who can see at a glance, feel in a moment, do the impossible, and know the unknown …

Human beings cannot live without challenge. We cannot live without meaning. Everything ever achieved we owe to this inexplicable urge to reach beyond our grasp, do the impossible, know the unknown. The Upanishads would say this urge is part of our evolutionary heritage, given to us for the ultimate adventure: to discover for certain who we are, what the universe is, and what is the significance of the brief drama of life and death we play out against the backdrop of eternity.

(Anonymous – The Upanishads)

Notes

1 This suggestion by Niebuhr is contradicted by myriad examples of communities which, by way of social mores and pressures, corral and guide the otherwise disruptive behaviour of recalcitrant individuals. While such communities exemplify a desirable form of governance which might possibly be scaled up, Niebuhr is referring to societies in which the appropriate cultural infrastructure is absent, and to societies which have grown so large that interpersonal communication, empathy, and cohesion are beyond reach.
2 William Rees, personal communication, with permission.
3 There is an unfortunate declension of value implicit in this metaphor, from gold to silver to bronze. Whether intended or not, this does not fit well with an egalitarian society in which all roles are of equal worth.
4 In his most recent book Thomas Homer-Dixon has developed a carefully nuanced discussion of hope, arguing it must be conditioned by a realistic assessment of where we are now (honest hope); by an ability to choose wisely among alternative paths forward (astute hope); and by the psychological wherewithal to drive/inspire personal/collective agency (powerful hope) (Homer-Dixon 2020).

References

Arendt, Hannah. 1958. *The Human Condition*. Chicago, IL: University of Chicago Press.
Aristotle. [1951]. *The Politics*. Translated by T. A. Sinclair. Harmondsworth, England: Penguin Books Ltd.
Gare, Arran E. 1995. *Postmodernism and the Environmental Crisis*. New York: Routledge.
Hamilton, Alexander, James Madison, and John Jay. [2010]. "Federalist #57." In *The Federalist Papers*. Seattle, WA: CreateSpace.
Havel, Vaclav. 1986. *The Politics of Hope*. https://en.wikiquote.org/wiki/V%C3%A1clav_Havel
Homer-Dixon, Thomas. 2020. *Commanding Hope: The Power We Have to Renew a World in Peril*. Toronto: Alfred A. Knopf Canada.
Niebuhr, Reinhold. [1932] 1960. *Moral Man and Immoral Society: A Study in Ethics and Politics*. New York: Charles Scribner's Sons.
Ophuls, William. 2011. *Plato's Revenge: Politics in the Age of Ecology*. Cambridge, MA: The MIT Press.
Passerin d'Entreves, Maurizio. 2019. "Hannah Arendt." In *The Stanford Encyclopedia of Philosophy*, Fall, edited by Edward N. Zalta. Metaphysics Research Lab, Stanford University. https://plato.stanford.edu/archives/fall2019/entries/arendt/

Plato. [1974]. *The Republic*. Translated by G.M.A. Grube. Indianapolis, IN: Hackett Publishing Company.

SustainAbility/Globescan. 2012. *The Regeneration Roadmap: Global Expert Perspectives on the State of Sustainable Development*. www.globescan.com/component/edocman/?view=document&id=8&Itemid=591

10 The End and the Beginning

This book does not address the myriad practical concerns shared by many with respect to securing the long-run relationship between people and planet. Instead, it is intended as a prologue to a formative, catalytic discourse about the implicate universe we inhabit, about our place on Earth, and about the clear and present existential dangers that threaten our tenure here. We have delayed too long, missed too many opportunities for change, and failed to come to terms with the prospect of grievous harm inflicted by ourselves on ourselves, and on the planet which gave us life. Notwithstanding the hard work and commitment of countless people around the world, the business-as-usual trajectory still dominates the global agenda, still informs the system of global governance which steers us recklessly and rapidly in the wrong direction.

The physical and biological sciences provide information and new tools for practical action, but the social sciences and humanities must also be more effectively engaged if we hope to change the underlying suite of attitudes, values, and worldviews which shape and give meaning to the modern human experience. Notwithstanding the better future promised by the story of Progress and Prosperity, today our shared mindset is uneasy. Our relationship with Earth is precarious. We have no clear sense of why we are here or where we might be going, and undercurrents of fear and fatalism are beginning to corrode the positive, hopeful sentiments inherent to human nature. Who will lead the way?

Agency now needs to be rethought, recalibrated, and rebuilt according to the principle of self-transcendence, meaning to see beyond ourselves as individuals, and to see beyond our own

DOI: 10.4324/9781032647067-12

species to the universal context of our shared existence. In this regard Vaclav Havel reminds us

> of what we have long projected into our forgotten myths and what perhaps has always lain dormant within us as archetypes. That is, the awareness of our being anchored in the Earth and the universe, the awareness that we are not here alone nor for ourselves alone ... This forgotten awareness is encoded in all religions. All cultures anticipate it in various forms. It is one of the things that form the basis of man's understanding of himself (sic), of his place in the world, and ultimately of the world as such ... This awareness endows us with the capacity for self-transcendence.
>
> (Havel 1994)

And, crucially, this unique capacity gives us privileged proximity to supra-human virtues which form the anchor and the foundation from which renewed moral evolution might proceed, and from which a new metanarrative might emerge – a new story featuring us as co-participants in a play whose meaning we will both discover and determine.

Deus ex Machina

The role and impact of powerful contemporary religious institutions notwithstanding, today's metanarrative of Progress and Prosperity is predominantly secular, leaving little room for theological musings. However, because bringing the humanities back into the conversation about our environmental emergency may be critical to our eventual success with respect to that challenge, I will pursue this line of thinking a little further.

I have referred thematically throughout this book to 'moral evolution,' 'absolute values,' even 'God,' in one fashion or another. I approach these topics reluctantly, as an unlikely escape route, and with considerable doubt, if not disdain. Nonetheless, I find the prospect of 'co-participating in a play whose meaning we will both discover and determine' intriguing. With this in mind, then, and in order to move the play along, I adopt here the position that the universe we see is a singular entity, hospitable to life, and that the explanation for its existence which calls upon divine creation is at least as credible as the notion of a multiverse in which everything that can happen, will happen. On this view, then, in the beginning a preternatural entity, existing in an undivided, singular form, might have willingly and voluntarily undergone a process of, in effect, disintegration into disparate parts, those parts being material (understood as matter/energy) and epiphenomenal (understood as information embodied in relationships between parts). Energy made this new universe dynamic; information ensured that change would not occur in a disorderly manner. These two universal and indestructible ontological pillars of reality form the foundations of all we can see and manipulate, including the synthetic evolution of all the familiar elements of the periodic table, yet their essence remains fundamentally elusive, even mysterious.

This narrative suggests that all forms of energy, all particulate matter, and all the laws of physics, are residual components of the originating entity. If that entity were 'alive' in any meaningful sense of the word, then all its fractured pieces would likewise carry with them the life force from which they emerged. This point of view is sometimes called 'pantheism,' essentially the argument that the universe – everything – is imbued with whatever characteristics were inherent to the originating entity. This supposition carries with it an important and, in my view, compelling explanatory advantage regarding the 'origin of life' question. The multiverse theory offers no explanation whatever for the fact that the elements as we know them – their atomic structure and the laws that govern their interactions – should be such that they can somehow coalesce to produce living organisms, especially us, with the unique ability to gaze reflectively on the universe. Through us the universe has become conscious of itself, which allows us the possibility of becoming co-participants in its evolution. Since the moment of the Big Bang, the universe has contained within itself the potential for carbon-based life forms – the magic of life has always been inherent to the nature of all that is – but the probabilistic, reductionist multiverse theory has nothing to say about why matter/energy, in whatever form, should have anything at all to do with the creation of life, or why consciousness should arise in such a self-reflective way. Pantheism, on the other hand, offers the simple idea that if God is alive, then so must the universe and all its disparate parts be. This leaves unanswered, however, the question of why the process of 'disintegration' should have been undertaken 'willingly and voluntarily' in the first place.

Science and Theology

Cosmologist/theologian John Polkinghorne has offered a useful template for a reconciliation of science with theology, at least with respect to the question of order. Much is known about the initial conditions of the universe because at that time those conditions were actually very simple. In fact, the native characteristics of the nugget from which the Big Bang erupted were so tightly constrained (to a precision of one part in more than 10^{123} according to Roger Penrose[1]) that the 'fine-tuning' of other physical constants, previously mentioned, pales by many orders of magnitude. In other words, the beginning of the universe was highly ordered with exceptionally low entropy.[2] The occurrence of the Big Bang, then, would require either a remarkably propitious set of conditions to pop into existence spontaneously, by chance, from a pre-existing quantum vacuum,[3] or it was the purposeful expression of a supernatural reality. If it was the latter, then the Big Bang was a moment of divine creation perpetrated, presumably, by God.

A recurring and deeply perplexing puzzle in all branches of theology has been the nature of the relationship between necessity and contingency – between, for example, the necessarily perfect Forms of the Platonic realm and the contingent details of actual but imperfect worldly events, and how one may proceed from the other. One wonders, if God is necessarily perfect, then why would the Big Bang, pregnant with the promise of carbon-based life but including the potential for harm as well as good, be allowed to occur? Polkinghorne answers this way:

> The act of creation is an act of divine self-limitation — an act of kenosis, as the theologians say — on the part of the Creator in allowing creatures truly to be themselves and to make themselves. This implies that, although allowed by God, not all that happens will be in accordance with positive divine will ... the Universe is not a divine puppet theatre ... evil and suffering are the inescapable shadow side of evolutionary fruitfulness.
>
> (Polkinghorne 2007b, 3)

In other words, creation is costly to God in the sense that some amount of perfection is given up, some degree of necessity is abandoned for the sake of encountering the living challenge of a contingent world. David Ward expresses a similar opinion by suggesting that Plato's world of archetypes is opened by God to the phenomena of the physical cosmos which participate in it "partially and imperfectly" (Ward 2010, 286). This overlap of types of being – the timeless and the merely temporal – allows God to influence and be influenced by the unfolding reality of the universe as we know it. The interplay of necessity and chance, the intertwining of degrees of order with an open sensitivity to change means that "The Universe is a world of true becoming in which the future is not an inevitable consequence of the past" (Polkinghorne 2007a, 4).

Atheistic or humanist science and those with a penchant for naturalism want no truck with such a deity. They turn instead to the physical and biological sciences for materials needed to construct a new origin story, a new cosmology which can accommodate the (evolutionary) emergence of value and meaning, but which will also inspire wonder and restore a deep concern for all planetary biota, with whom we share a common ancestor. Because this alternative narrative is ostensibly evidence based, it is believed to be true and universal, so proponents of it argue that 'epic science' can provide sound information and the motivation necessary to galvanize widespread action on environmental issues.[4]

Stuart A. Kauffman develops this perspective at length by embracing the concept of emergence in complex systems. Emergence posits the existence of whole, self-organized structures which possess causal power[5] above and beyond the lawful constraints imposed on the constituents of elementary matter. These structures occur at all levels of reality, both physical and social, from the molecular to the institutional, and they interact with each other in ways that involve a certain degree of uncertainty because of their autonomous interaction with the determinative laws of physics. It is this emergent uncertainty in combination with quantum fuzziness that produces the 'endless creativity of nature,' which Kauffman tells us is the scientific stand-in for God. Moreover, not only is nature endlessly creative (that is, open to novelty), but evolution can also account for the appearance of agency, interpretation, values, meaning, and purpose. Kauffman argues that these concepts appear prototypically even in early life forms such as bacteria which can act (agency), read environmental signals pertaining to the availability of food (interpretation), appreciate the importance of those signals (valuation, and therefore meaning), and move to accomplish the objective of consumption (purpose). On this view, there is no need to call upon theology to explain these features of life which finally appear in their fully evolved form with human consciousness and will.

This point of view satisfies the secular predisposition to explain rather than understand. Pursuing this debate in detail would take us too far afield, but it is worth noting that Kauffman's argument in some ways misplays both science and theology. His bacteria example above strikes me as just another form of materialist reductionism, and his atheism prompts him to misleadingly portray God, if such an entity were to exist, as imposing clear strictures of moral behaviour on humanity, thus delimiting the freedom to choose, and simultaneously arming fundamentalists with sets of beliefs which could only lead to conflict. The 'epic science' argument, in other words, is as dependent on subjective interpretations and leaps of speculation as any kind of intuition. In short, neither science nor theology can present a logically coercive argument, so it may be that choosing between the two is a matter of personal taste. In this light and given the seriousness of the environmental emergency and the existential threat it presents, it might be prudent to open the door to the humanities a little wider.

The theological discussion presented here (pantheism) postulates that the phenomena of the physical universe – in other words, reality as we perceive it – are an immanent manifestation of God and an open-ended encounter with life in which we all participate. It suggests that the universe is a co-evolutionary partnership between God and life forms such as ourselves who can lead lives of our own choosing yet still appreciate that part of creation still unblemished by contingencies, that is, to the perfection of God still at play as a necessary precondition for reality to exist at all. But that perfection is vulnerable, vulnerability being the cost of admission to the adventure of real life. And here we find an answer to the question posed much earlier: why do we have the capacity to perceive and understand that which constitutes the universe itself, its underlying laws and processes, when this capacity serves no survival or reproductive purpose? The answer may be that we are co-evolutionists endowed with the responsibility to participate in an interaction between the imperfect and the perfect, the contingent and the necessary, the timeless and the merely temporal. To what purpose I leave the reader to speculate, but clearly meeting this teleological impulse in whatever form would not be possible without the cognitive and transcendent abilities we have and, of course, it remains to be seen how adeptly we use those tools. If we act as best as our natures permit, presumably we will survive and, with good fortune, flourish. If not, the worldline of all that is may be permanently ended.

Empirical, reductionist science is not helpful with respect to the moral evolution of modern society. It is prosaic and, without access to the poetry of life or to the transcendent dimension of reality, it falls prey to the foibles which afflict any kind of materialist worldview and, ultimately, it founders on the hard rocks of humanity's hubris. We are told by evolutionary science that our concepts of truth, morality, even beauty emerged from some kind of evolutionary imperative, that they serve useful social and reproductive purposes. From an intuitive or revelatory point of view, this is, seemingly, a vain and wilful exaggeration of human exceptionalism. Instead, our capacity to understand as well as to explain the world around makes us full participants in a self-aware, co-evolving universe, and our unique ability to perceive and partake of the intrinsic reality of the Good, the True, and the Beautiful provides the opportunity and the relational context from which we may find our bearings in the pursuit of a morally felicitous future.

Do Not Go Gentle

Among the elderly, conversation often drifts to matters of health. Ailments are described, treatments discussed, prognoses hopefully compared. These conversations happen because the various medical experiences of the old and infirm tend to be top of mind, and because the cumulative effect of them suggests the inevitability of a casket, which puts a sharper edge on things. It is not a question of losing faith, or hope, or one's verve for life; dealing with health issues is simply a matter of prudential care, and to some extent a matter of academic interest in the process of waning physical vitality. For some, these conversations serve no useful purpose other than as something to push off from the better to squeeze the very most out of the remainder of life. Dylan Thomas captures this sentiment in his well-known poem:

> Do not go gentle into that good night,
> Old age should burn and rave at close of day;
> Rage, rage against the dying of the light.[6]

This is sage advice, and a nice remedy for the kind of morbidity which some poor souls may fall prey to. It does not recommend a happy forbearance and it certainly does not advocate a comfortable slide into dotage; it calls for a rebellion against closure, a refusal to go quietly, a demand to be allowed to continue to live. It is a call to arms against a process already deemed decided.

It is tempting to apply the energy and enthusiasm aroused by Thomas's poem to the problem of planetary environmental degradation, about which a positive, damn-the-torpedoes frame of mind is also called for. And in fact this attitude is common among activists, and many others who take environmental issues seriously, simply expressed as "We know we're in trouble but it's not too late, solutions are available, so let's roll up our sleeves and get to work!" This is the popular slogan of the practical school of change. Motivated people display a steely optimism, a determined conviction that with hard work things will turn out well. But this indomitable approach to change is not actually compatible with Dylan Thomas's poem, which illuminates the harder question of how to respond to the inevitability of death. Things do not necessarily turn out well. His point is to face what is unavoidable with fierce energy, with all remaining strength, to challenge it and in that way to exact a kind of triumph over something beyond our control, knowing all the while that "Wise men at their end know dark is right," as Thomas writes in his second verse.

But of course, these admirable phrases apply to an aging person, not a degrading biosphere, which is not the same thing. For the latter, is it really the close of day? Is the light really dying? Earth is not trending to death, even though the extinction of many thousands of species at the careless hands of *Homo sapiens* is certainly death to those whose termination has been effected. Barring the worst case in which Earth becomes permanently and completely uninhabitable, the biosphere will not die; it will rebound and teem once again with life, albeit after a recovery period of many millions of years. In what sense, then, might this poetic verse be relevant to planetary degradation?

The relevance begins from the observation that the human/nature complex – that is, the socio-ecological complex – is indivisible. We are irrevocably embedded in our natural surroundings, so if the environment degrades, we will too. This is simply a matter of fact, but it takes on an interesting moral cast with the additional point that our species is responsible for that degradation which inflicts harm not only on Earth's ecosystem as such, but on our own well-being and survival chances. In that foreboding context, 'the dying of the light' could be read to refer to the precipitous weakening of the vitality of Earth and, simultaneously, to the reduced possibilities available for us to explore and realize our full potential.

Nor, however, can an aging person be precisely compared with society in general – to the human population at large – because the former will surely die even as the latter lives on, perhaps even in spite of itself. Human society on Earth may never die. And herein lies the still-open possibility that, unlike the individual but like the biosphere, society might rebound from decay – might learn lessons from errors of judgement and behaviour, might even achieve a higher level of social consciousness replete with benefits, both material and psychic, never available to the now departed individual person. Humanity might grow up, and this surely is something worth fighting for, losses notwithstanding. Thomas's poem, then, calls us to rave and rage not against the death of Earth's biosphere, or the death of human society on Earth, but against the inevitable loss of our naïveté, our youthfulness, and the carefree immaturity which has brought us to the edge of environmental tragedy.

I claimed the world and made it mine.
I prospered abundantly.

But the venality of pride
The artlessness of adolescence

And the foreboding of my future
Confused me.
I do not know who I am.

I do not know who I am but I do know this.
The world is mine and I will use it as I please.
It pleases me now and I will use it.
If all things end it will be no concern of mine
That all things have ended.
If I did not end I could not understand
The meaning of an endless world.

Notes

1 Only a very special set of initial conditions, plus specific laws, can explain the universe as we see it today. See Penrose (1989, 339–45).
2 Entropy is a statistically precise measure of disorder.
3 And if it happened once, it could happen any number of times in an infinite, ageless quantum ocean, hence the origin of the multiverse theory.
4 For more on this, see the discussion "Cosmology and the Environment" at *The Immanent Frame: Secularism, Religion and the Public Sphere.* https://tif.ssrc.org/2015/09/14/cosmology-and-the-environment/
5 To be clear, 'causal power' here refers to the imposition of patterns and determinative or guiding influences which emanate from the whole organization of emergent structures, sometimes called 'downward causation'. These patterns and influences do not constitute new forces beyond the four known forces of nature.
6 The full poem, a 19-line villanelle, is available at http://poets.org/poem/do-not-go-gentle-good-night/

References

Havel, Vaclav. 1994. "Acceptance Speech for Liberty Medal, Philadelphia, USA," July 4. https://constitutioncenter.org/libertymedal/recipient_1994_speech.html

Penrose, Roger. 1989. *The Emperor's New Mind: Concerning Computers, Minds and the Laws of Physics.* Oxford: Oxford University Press.

Polkinghorne, John. 2007a. *The Anthropic Principle and the Science and Religion Debate,* Faraday Paper #4, April. Cambridge: The Faraday Institute for Science and Religion, St. Edmund's College, UK.

Polkinghorne, John. 2007b. *The Science and Religion Debate: An Introduction*, Faraday Paper #1, April. Cambridge: The Faraday Institute for Science and Religion, St. Edmund's College, UK.

Ward, Keith. 2010. "God as the Ultimate Information Principle." In *Information and the Nature of Reality: From Physics to Metaphysics*, edited by Paul Davies and Niels Henrik Gregersen, 282–300. Cambridge: Cambridge University Press.

Index

Note: Endnotes are indicated by the page number followed by "n" and the note number e.g., 36n4 refers to note 4 on page 36.

For Product Safety Concerns and Information please contact our EU
representative GPSR@taylorandfrancis.com
Taylor & Francis Verlag GmbH, Kaufingerstraße 24, 80331 München, Germany